QUAKES, ERUPTIONS AND OTHER GEOLOGIC CATACLYSMS

QUAKES, ERUPTIONS AND OTHER GEOLOGIC CATACLYSMS

The Changing Earth Series

JON ERICKSON

Facts On File®

AN INFOBASE HOLDINGS COMPANY

94-996

QUAKES, ERUPTIONS AND OTHER GEOLOGIC CATACLYSMS

Facts On File, Inc.
460 Park Avenue South
New York NY 10016

Library of Congress Cataloging-in-Publication Data
Erickson, Jon, 1948–
   Quakes, eruptions, and other geologic cataclysms / Jon Erickson.
     p. cm. — (Changing earth)
   Includes bibliographical references and index.
   ISBN 0–8160–2949–0 (alk. paper)
   1. Natural disasters. I. Title. II. Series: Erickson, Jon,
  1948– Changing earth.
  GB5014.E75 1994
  550—dc20   93–43587

Facts On File books are available at special discounts when purchased in bulk quantities for businesses, associations, institutions or sales promotions. Please call our Special Sales Department in New York at 212/683–2244 or 800/322–8755.

Text design by Ron Monteleone/Layout by Robert Yaffe
Jacket design by Catherine Hyman
Printed in the United States of America

RRD FOF 10 9 8 7 6 5 4 3 2 1

This book is printed on acid-free paper.

CONTENTS

TABLES vii

ACKNOWLEDGMENTS viii

INTRODUCTION ix

1
THE DYNAMIC EARTH 1
THE NEW GEOLOGY • MANTLE CONVECTION • SEAFLOOR SPREADING •
SUBDUCTION ZONES • PLATE INTERACTIONS

2
EARTHQUAKES 16
MAJOR QUAKES • AREAS AT RISK • EARTHQUAKE FAULTS • EARTHQUAKE
CAUSES • EARTHQUAKE DAMAGE • SEISMIC SEA WAVES

3
VOLCANIC ERUPTIONS 35
VIOLENT VOLCANOES • THE FIRE BELOW • VOLCANIC ACTIVITY • GAS
EXPLOSIONS • HAZARDOUS VOLCANOES

4
EARTH MOVEMENTS 52
LANDSLIDES • ROCKSLIDES • SOIL SLIDES • MUDFLOWS • SUBMARINE
SLIDES • SOIL EROSION

5
CATASTROPHIC COLLAPSE 68
THE SINKING EARTH • GROUND FAILURES • SUBSIDENCE • RESURGENT
CALDERAS • COLLAPSE STRUCTURES

6
FLOODS 84
HAZARDOUS FLOODS • THE HYDROLOGIC CYCLE • HYDROLOGIC MAPPING •
FLOOD-PRONE AREAS • FLOOD TYPES • DRAINAGE BASINS • FLOOD
CONTROL

7
DUST STORMS 102
DESERT REGIONS • DESERTIFICATION • HABOOBS • SAND DUNES •
DUST BOWLS

8
GLACIERS 119
THE POLAR ICE CAPS • CONTINENTAL GLACIERS • CAUSES OF GLACIATION
• GLACIAL SURGE • RISING SEA LEVELS

9
IMPACT CRATERING 133
CRATERING EVENTS • CRATERING RATES • METEORITE IMPACTS • ROGUE
ASTEROIDS • STONES FROM THE SKY • IMPACT EFFECTS

10
MASS EXTINCTIONS 149
HISTORIC EXTINCTIONS • CAUSES OF EXTINCTIONS • EFFECTS OF
EXTINCTIONS • MODERN EXTINCTIONS • THE WORLD AFTER

GLOSSARY 167

BIBLIOGRAPHY 179

INDEX 186

TABLES IN QUAKES, ERUPTIONS AND OTHER GEOLOGIC CATACLYSMS

1
THE 10 MOST LETHAL EARTHQUAKES SINCE 1900   16

2
GREATEST VOLCANIC ERUPTIONS: TNT EQUIVALENT AND MATTER EJECTED   35

3
THE MOST DESTRUCTIVE U.S. FLOODS   84

4
ALBEDO OF VARIOUS SURFACES   106

5
CHRONOLOGY OF THE MAJOR ICE AGES   125

6
RADIATION AND EXTINCTION FOR MAJOR ORGANISMS   149

7
FLOOD BASALT VOLCANISM AND MASS EXTINCTIONS   156

8
COMPARISON OF MAGNETIC REVERSALS WITH OTHER PHENOMENA   158

ACKNOWLEDGMENTS

The author thanks the following organizations for providing photographs for this book: the National Aeronautics and Space Administration (NASA), the National Oceanographic and Atmospheric Administration (NOAA), the National Park Service, the U.S. Air Force, the U.S. Army Corps of Engineers, the U.S. Department of Agriculture–Forest Service, the U.S. Department of Agriculture–Soil Conservation Service, the U.S. Geological Survey (USGS), and the U.S. Navy.

INTRODUCTION

Geologic hazards have plagued people since time immemorial. This is because we live on a dynamic planet, with devastating earthquakes, disastrous volcanoes, and other catastrophic geologic activities that destroy property and take human lives. These phenomena arise from the interactions of a jumble of crustal plates that comprise the Earth's outer shell. Tectonic forces are also responsible for raising mountains and creating a variety of geologic structures, often accompanied by powerful earthquakes.

Earthquakes are the most destructive natural force. Massive earthquakes produce widespread damage, destroying entire cities and killing people by the thousands. A large portion of the land surface is crisscrossed by active faults, mostly on plate boundaries at the edges of continents. Since half the world's population lives in coastal regions, it is placed at great risk from earthquake destruction. Tsunamis generated by undersea earthquakes and powerful volcanic eruptions are particularly hazardous to coastal inhabitants.

Volcanoes are the next most destructive natural force. Since the dawn of man, thousands of volcanoes have erupted, nearly two-thirds of which have caused fatalities. During the past 400 years, over 500 volcanoes have erupted, killing over 200,000 people and destroying billions of dollars worth of property. Presently, some 600 active volcanoes exist throughout the world that are poised for eruption at any time. As the world's population continues to grow and people insist on living in the domain of volcanoes, future eruptions could be far more deadly.

Other geologic hazards include ground failures, floods, and dust storms. The ground can give way even on the gentlest slopes. Material on most slopes is constantly on the move, from the slow creep of soil and rock to catastrophic landslides and rockfalls that travel at tremendous speeds. All earth movements are naturally recurring events that have become increasingly hazardous because of human activities. Some ground failures are due to the dissolution of soluble materials underground or the withdrawal of fluids from subsurface sediments. Other ground failures result when subterranean sediments liquefy during earthquakes or violent volcanic eruptions, causing considerable damage to man-made structures.

Floods are naturally occurring events that become hazardous only when people build on floodplains, whose purpose is to carry away excess water during river overflows. Failure to recognize this function has led to a rapid and haphazard development of floodplains and a consequent increase of flood risks. Floodplains are a valuable natural resource that must be managed properly to prevent flood damage. Unfortunately, improper use of floodplains has led to serious damage to property and loss of life. Dam-break floods are few but significantly dangerous, especially if dams fail due to geologic hazards such as earthquakes and landslides.

Dust storms can directly threaten life, and both people and animals have died of suffocation during severe storms. Another direct threat of dust storms to man is soil erosion. Soil is an unrenewable natural resource that is disappearing globally at alarming rates. Many countries are at risk from soil erosion and advancing deserts, undermining efforts for the world's population to feed itself. Furthermore, much of the Earth's surface is becoming desertified because of poor stewardship of the land, including deforestation and improper agricultural practices.

The melting of the world's great glaciers during a sustained warmer climate could raise sea levels and drown coastal regions. Beaches and barrier islands would disappear as the shoreline continues to advance inland. Without protective barriers, raging storms continually batter the seashores. The rising waters would inundate rich river deltas that feed a large portion of the world's population. Coastal cities would have to move farther inland or erect costly seawalls to protect against the rising sea.

The Earth is constantly being pelted by meteors, and meteorite falls are more common than people think, as many examples of meteorites crashing into houses and cars have shown. A large number of asteroids cross the Earth's path around the sun, and a major impact could be catastrophic to all life. Yet it is remarkable how life has managed to survive, considering all the great upheavals throughout geologic history. The planet has been subjected to a number of major catastrophes, resulting in the extinction of large numbers of species. Over 99 percent of all species that have inhabited the Earth over geologic time have gone extinct. Present-day extinctions, however, are caused by our destructive activities. As human populations continue to expand out of control, other species are forced aside to make room for many more of us. This places humans in the unique position of being among the most destructive forces on Earth.

1

THE DYNAMIC EARTH

The Earth is a highly dynamic planet, with rising mountain ranges, gaping canyons, erupting volcanoes, and shattering earthquakes. No other body in the Solar System offers so many unusual landscapes, sculpted by highly active weathering agents that can cut down tall mountains and gouge out deep ravines. These activities are expressions of plate tectonics (Fig. 1), a force that moves the continents around the globe, thus making the Earth a living planet both geologically and biologically. If it were not for a jumble of crustal plates interacting with each other to provide us with a myriad of geologic features, ours would be a desolate world.

THE NEW GEOLOGY

The Earth's outer layer is broken into seven major and several minor lithospheric plates (Fig. 2), composed of the crust and the upper brittle mantle called the lithosphere. The lithospheric plates ride on a hot pliable layer, called the asthenosphere, and their constant interaction with each other shapes the surface of the planet. The shifting plates range in size from a few hundred to millions of square miles and average about 60 miles thick. The lithosphere rides freely on the semimolten layer of the upper mantle.

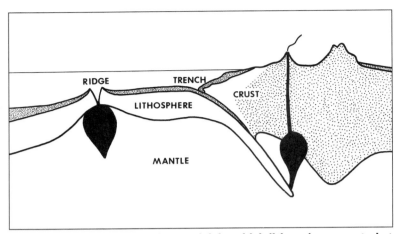

Figure 1   The plate tectonics model, in which lithosphere created at midocean ridges is subducted into the mantle at deep-sea trenches, causing the continents to drift around the Earth.

This structure is unique and important for the operation of plate tectonics, which is responsible for the Earth's active geology.

These lithospheric plates meet at two different kinds of intersections: divergent plate margins, where the plates are pulling away from each other, and convergent plate margins, where they collide. The divergent plate margins are spreading ridges on the deep ocean floor, where basalt welling up from within the mantle creates new oceanic crust in a process known as seafloor spreading. The midocean ridge system snakes around the globe like the stitching on a baseball for a distance of 40,000 miles, thus making it the longest structure on Earth. Some of the molten magma erupts on the surface of the ridge as lava, but the vast majority cools and bonds to the edges of separating plates. Periodically, however, molten rock overflows onto the ocean floor in gigantic eruptions, providing several square miles of new oceanic crust yearly.

The convergent plate margins are deep-sea trenches, called subduction zones, where old oceanic crust sinks into the mantle to provide new basalt in a continuous cycle of crustal generation. If tied end to end, the subduction zones would stretch clear around the world. The convergence rates between plates range from less than 1 inch to over 4 inches per year, corresponding to the rates of plate divergence.

Since the Earth is not expanding, new oceanic crust created at spreading ridges must be destroyed in subduction zones. Most subduction zones surround the Pacific Basin. Plate subduction is responsible for the intense seismic activity that fringes the Pacific Ocean in a region known as the circum-Pacific belt. Furthermore, wide bands of earthquakes mark convergent continental plate margins, whereas narrow bands of earthquakes mark many major oceanic plate boundaries. The circum-Pacific belt is synonymous with the Ring of Fire, known for its extensive volcanic activity. Subduction zone volcanoes form island arcs, mostly in the Pacific, and most volcanic mountain ranges on the continents.

The lithospheric plates carry continental crust around the surface of the Earth like ships frozen in arctic pack ice. The bulk of the crust comprises granitic and metamorphic rocks, which constitute most of the continents. Because the continental crust contains light materials, it remains on the surface due to its greater buoyancy. When two plates collide, the less buoyant oceanic crust subducts under continental crust or younger oceanic crust. As the oceanic crust descends into the mantle, it remelts to provide new molten magma. The line of subduction is marked by the deepest trenches in the world.

Plate collisions uplift mountain ranges on the continents and create volcanic islands on the ocean floor. When an oceanic plate subducts beneath a continental plate, it forms sinuous mountain chains, like the Andes of South America, and volcanic mountain ranges, like the Cascades of the American Pacific Northwest (Fig. 3). The breakup of a plate creates new continents and oceans, and the collision of plates builds supercontinents. The rifting and patching of the continents has been an ongoing process for at least 2.7 billion years and possibly even longer.

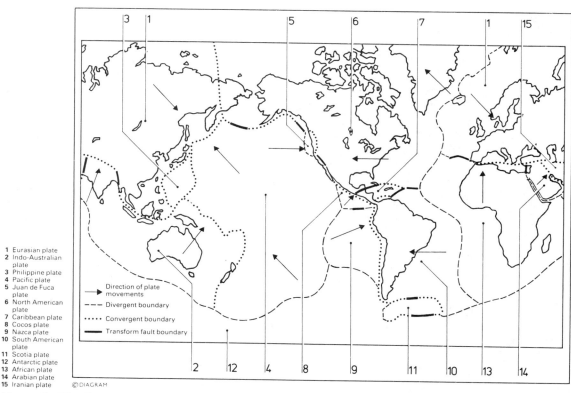

1 Eurasian plate
2 Indo-Australian plate
3 Philippine plate
4 Pacific plate
5 Juan de Fuca plate
6 North American plate
7 Caribbean plate
8 Cocos plate
9 Nazca plate
10 South American plate
11 Scotia plate
12 Antarctic plate
13 African plate
14 Arabian plate
15 Iranian plate

→ Direction of plate movements
– – – Divergent boundary
· · · · Convergent boundary
—— Transform fault boundary

©DIAGRAM

Figure 2   The Earth's crust is fashioned out of several lithospheric plates that are responsible for the planet's active geology.

The continents most recently rifted apart and drifted away from each other about 180 million years ago (Fig. 4), when a supercontinent called Pangaea (from Greek roots meaning all lands) rifted apart along the present Mid-Atlantic Ridge. Upwelling magma created an undersea volcanic mountain range that runs through the middle of the Atlantic Ocean, weaving halfway between the continents surrounding the Atlantic Basin. The Mid-Atlantic Ridge is part of a global spreading ridge system that is responsible for creating new oceanic crust. Molten magma rising from deep within the mantle forms new lithosphere by the addition of basalt to plate margins.

The two sides of the Atlantic Basin spread apart an inch or more per year. As the Atlantic Basin widens, the surrounding continents separate, compressing the Pacific Basin. The Pacific Basin is ringed by subduction zones that assimilate old lithosphere, causing the Pacific plate along with adjacent plates to shrink in size. The Pacific plate, the world's largest lithospheric plate, originated as a microplate no bigger than the United States some 190 million years ago. It achieved its present size by the gradual addition of lithosphere at associated spreading ridges.

The oceanic crust, composed of basalts originating at spreading ridges and sediments washed off nearby continents, gradually increases density and eventually subducts into the mantle. On its journey deep into the Earth's interior, the lithosphere and its overlying sediments melt. The molten magma later rises toward the surface in huge bubblelike structures called diapirs, from the Greek meaning "to pierce." When the magma reaches the base of the crust, it provides new molten rock for magma chambers beneath volcanoes and granitic bodies called plutons, which often form moun-

Figure 3 The south side of Mount Adams, with Mount Rainier in the background, Cascade Range, Yakima County, Washington. Photo by Austin Post, courtesy of USGS

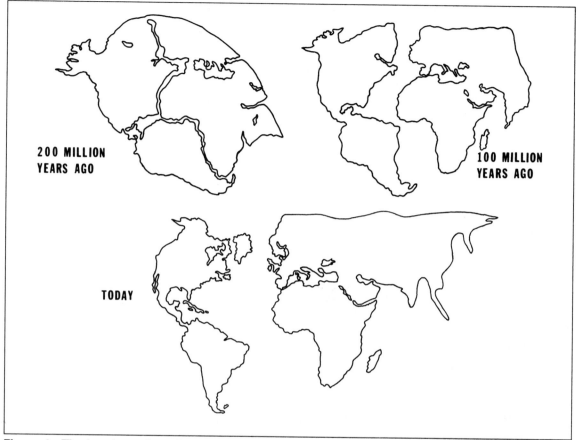

Figure 4   The breakup of continents. About 180 million years ago a supercontinent called Pangaea rifted apart to form the present-day continents.

tains. In this manner, plate tectonics is continuously changing and rearranging the face of the Earth.

MANTLE CONVECTION

All geologic activity taking place on the surface of the planet is an outward expression of a great heat engine in the Earth's interior (Fig. 5). The heat cycle within the mantle is the main driving force behind plate tectonics. Convection currents and mantle plumes, or columns, transport heat from the core to the underside of the lithosphere. This process is responsible for most of the volcanic activity on the ocean floor and on the continents. Most mantle plumes originate from within the mantle, while some arise from

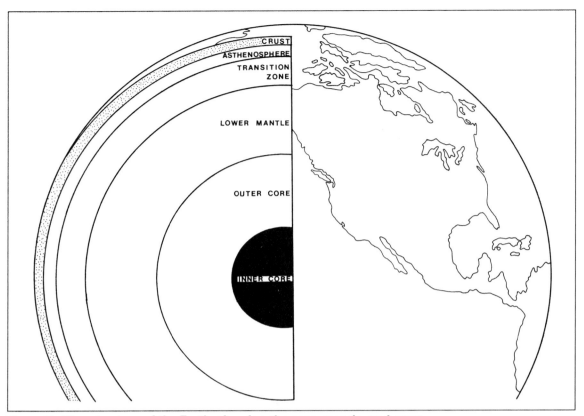

Figure 5   The structure of the Earth, showing the core, mantle, and crust.

the very bottom of the mantle, like huge bubbles rising from above the Earth's core.

Convection is the motion within a fluid medium resulting from a difference in temperature from top to bottom. Fluid rocks in the mantle acquire heat from the core, ascend, dissipate heat to the lithosphere, cool, and descend back to the core for more heat. Lithospheric plates, created at spreading ridges and destroyed at subduction zones, are the products of convection currents in the mantle.

The formation of molten rock in the mantle and the rise of magma to the surface is due to an exchange of heat within the planet's interior. The Earth is steadily losing heat from the interior to the surface through its outer shell, or lithosphere. Some 70 percent of this heat loss results from seafloor spreading, and most of the rest is due to volcanism at subduction zones. However, volcanic eruptions represent only localized and highly spectacular releases of this heat energy.

Most of the Earth's thermal energy is generated by radioactive isotopes, mainly potassium (K), uranium (U), and thorium (T), or simply KUT. The temperature within the Earth increases rapidly with depth. At a depth of about 70 miles, where the material of the upper mantle begins to melt, the temperature is about 1,200 degrees Celsius. This marks the semimolten region of the upper mantle, or the asthenosphere, on which the rigid lithospheric plates ride. The asthenosphere is constantly losing material, which adheres to the undersides of lithospheric plates. If the asthenosphere were not continuously fed new material from mantle plumes, the plates would grind to a halt, and the Earth would become, in all respects, a dead planet. Fortunately, this event is not expected for several billion years.

The temperature of the upper mantle increases gradually to about 2,000 degrees Celsius at a depth of 300 miles, and then rapidly increases to the top of the core, where temperatures reach 5,000 degrees. Most of the mantle's heat is generated internally by radiogenic sources. The rest is supplied by the core, which has retained much of its heat since the early accretion of the Earth some 4.6 billion years ago. The temperature difference between the mantle and the core approaches 1,000 degrees. Material from the mantle might mix with the fluid outer core to form a distinct layer on its surface that could block heat flowing from the core to the mantle and interfere with mantle convection.

The surface of the Earth is continuously being shaped by the action of the mantle churning over below the lithosphere (Fig. 6). The mantle currents travel very slowly, completing a single convection loop in perhaps hundreds of millions of years. Without mantle convection, erosion would wear down mountains to the level of the prevailing plain in a matter of only 100 million years, or merely 2 percent of the Earth's age. The surface of the Earth would then become a vast, featureless plain, unbroken by mountains and valleys. No volcanoes would

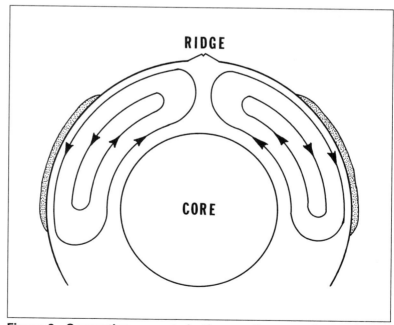

Figure 6   Convection currents in the mantle move the continents around the Earth.

erupt and no earthquakes would rumble across the land, leaving the planet as geologically and biologically dead as its moon.

The mantle convection cells act like rollers beneath a conveyor belt to propel the lithospheric plates forward. Hot material rises from within the mantle and circulates horizontally near the Earth's surface, where the top layer cools to form the rigid lithospheric plates that carry the crust around the surface of the planet. The plates complete the mantle convection by plunging back into the Earth's interior. In this manner, they are simply surface expressions of mantle convection. If fractures or zones of weakness appear in the lithosphere, the convection currents spread the fissures wider apart to form rift systems. This is where the Earth loses the largest portion of its interior heat to the surface, as magma flows out of the rift zones to form new oceanic crust.

SEAFLOOR SPREADING

Seafloor spreading, which creates new lithosphere at spreading ridges on the ocean floor, generates more than half the Earth's crust. Seafloor spreading begins with hot rocks rising by convection currents in the mantle. After reaching the underside of the lithosphere, the mantle rocks spread out laterally, cool, and descend back into the Earth's interior. The constant pressure against the bottom of the lithosphere fractures the plate and weakens it.

As the convection currents flow outward on either side of the fracture, they carry the separated parts of the lithosphere along with them, widening the gap. The rifting reduces the pressure, allowing mantle rocks to melt and rise through the fracture zone. The molten rock passes through the lithosphere and forms magma chambers that supply molten rock for the generation of new lithosphere. The greater the supply of magma to the chambers the higher the overlying spreading ridge is elevated.

The magma flows outward from a trough between ridge crests and adds new layers of basalt to both sides of the spreading ridge, thereby creating new lithosphere. The continents are carried passively on the lithospheric plates. Therefore, the engine that drives the birth and evolution of rifts and consequently the breakup of continents and the formation of oceans ultimately originates in the mantle.

The mantle rocks beneath spreading ridges that create new lithosphere consist mostly of peridotite, an iron-magnesium silicate. As the peridotite melts on its journey through the lithosphere, a portion becomes highly fluid basalt, the most common magma erupted on the Earth's surface. About 5 cubic miles of new basalt adds to the crust yearly, mostly on the ocean floor at spreading ridges. The rest contributes to the continued growth of the continents.

The spreading ridges are centers of intense seismic and volcanic activity, which manifests itself as a high heat flow from the Earth's interior. The greater the flow of magma to the ridge crest the more rapid the seafloor spreading. The spreading ridges in the Pacific Ocean are more active than those in the Atlantic and have less relief. Rapid spreading ridges do not achieve the heights of slower ones because magma has less opportunity to pile up into tall heaps like those in the Atlantic.

The spreading ridge system does not form a continuous line but is broken into small, straight sections called spreading centers (Fig. 7). The movement of new lithosphere generated at the spreading centers produces a series of fracture zones, which are long, narrow linear regions up to 40 miles wide that consist of irregular ridges and valleys aligned in a stairstep shape. When lithospheric plates slide past each other as the seafloor spreads apart, they create transform faults that range in size from a few miles to several hundred miles long. The faults occur every 20 to 60 miles along the Mid-Atlantic Ridge. Transform faults seem to result from lateral strain, and are the predicted result of plate movement on the surface of a sphere. This activity appears to be more intense in the Atlantic, where the spreading ridge system is steeper and more jagged than in the Pacific and Indian oceans.

Transform faults of the Mid-Atlantic Ridge generally have more relief than those of the 6,000-mile-long East Pacific Rise, which is the great spreading ridge system of the Pacific and counterpart to the Atlantic ridge system. Fewer widely spaced transform faults exist along the East Pacific Rise, which marks the boundary between the Pacific and Cocos plates. Furthermore, the rate of seafloor spreading is 5 to 10 times faster for the East Pacific Rise than for the Mid-Atlantic Ridge.

SUBDUCTION ZONES

The spreading seafloor in the Atlantic is offset by the shrinking of the seafloor in the Pacific. Deep trenches ring the Pacific Basin where old lithosphere subducts into the mantle. These subduction zones, lying

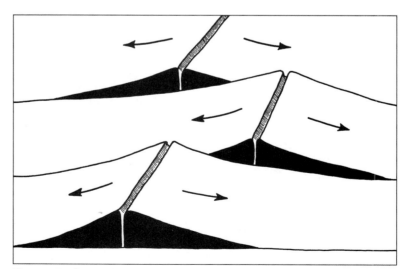

Figure 7   Spreading centers on the ocean floor are separated by transform faults.

off continental margins and adjacent to island arcs, are regions of intense volcanic activity that produce some of the most explosive volcanoes on Earth. The volcanic island arcs that fringe the subduction zones share a similar curved pattern; this curve is the geometric shape that develops when a plane cuts a sphere, for example when a rigid lithospheric plate subducts into the mantle.

The subduction zones are also sites of almost continuous seismic activity deep in the bowels of the Earth. A band of earthquakes marks the boundary of a sinking lithospheric plate. As plates press against each other along dipping fault planes, they create compressional earthquakes that can be highly destructive. Such earthquakes have always plagued Japan and the Philippines as well as other islands connected with subduction zones.

Lithospheric subduction plays a fundamental role in plate tectonics and accounts for many of the geologic processes that shape the planet. As a plate cools, it grows thicker and denser by a process known as underplating, in which magma from the asthenosphere adheres to the underside of the plate. When the plate becomes thick and heavy, it loses buoyancy and sinks into the mantle at clearly defined subduction zones. As the plate descends, it drags the rest of the plate along with it somewhat like a locomotive pulling a freight train. Therefore, plate subduction is the main force behind plate tectonics, and the pull at subduction zones is favored over the push of spreading ridges to drive the continents around the surface of the globe.

Plate subduction in the Pacific forms some of the deepest trenches in the world. As a lithospheric plate extends away from its place of origin at a spreading ridge, it cools and thickens as more material from the asthenosphere sticks to its underside. Eventually, the plate becomes so dense it loses buoyancy and sags deeper into the mantle. The depth at which a lithospheric plate sinks as it moves away from a spreading ridge varies with its age. The older the lithosphere, the more basalt has adhered and the deeper the plate.

As the lithospheric plate sinks into the mantle, the line of subduction creates a deep trench that accumulates large amounts of sediment derived from an adjacent continent. The continental shelf and slope contain thick deposits of sediment washed off the nearby continent. When the sediments and their seawater content are caught between a subducting oceanic plate and an overriding continental plate, they are subjected to strong deformation, shearing, heating, and metamorphism (recrystallization without melting). The sediments are carried deep into the mantle, where they melt to become the source of new magma for volcanoes that fringe the subduction zones.

When the magma reaches the surface, it erupts on the ocean floor, creating new volcanic islands. Most volcanoes, however, do not rise above sea level and instead became isolated undersea volcanoes called seamounts. The Pacific Basin is more volcanically active and has a higher density of seamounts than the Atlantic or Indian basins. The number of undersea volcanoes increases with increasing crustal age and thickness. The tallest seamounts rise over 2.5 miles above the seafloor and exist in the western Pacific near the Philippine Trench, where the oceanic crust is more than 100 million years old. The average density of Pacific seamounts is up to 10 volcanoes per 5,000 square miles of seafloor, a considerably larger number of volcanoes than on the continents.

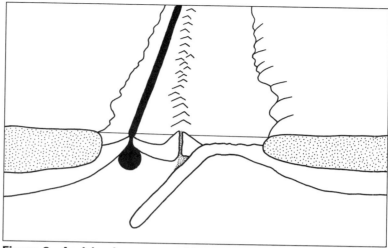

Figure 8   An island arc, created by a subduction zone where two plates converge, is caught between two colliding continents.

Subduction zone volcanoes are highly explosive because their magmas contain a large quantity of volatiles and gases that escape violently when reaching the surface. The type of volcanic rock erupted in this manner is called andesite, for the Andes Mountains that form the spine of South America and are well known for their explosive eruptions.

The seaward boundaries of the subduction zones are marked by deep trenches, which lie at the edges of continents or along volcanic island arcs (Fig. 8). Subduction zones, where cool, dense lithospheric plates dive into the mantle, are regions of low heat flow and high gravity. Conversely, the associated island arcs are regions of high heat flow and low gravity due to volcanism.

Behind the island arcs are marginal or back-arc basins that form depressions in the ocean floor due to plate subduction. Like the island arcs, back-arc basins are also regions of high heat flow because of the upwelling of magma from deep-seated sources. Deep subduction zones, like the Marianas Trench in the western Pacific, form back-arc basins. The Marianas Trench, which reaches a depth of nearly 7 miles, is the world's deepest trench and forms a long line northward from the island of Guam. Shallow subduction zones, like the Chilean Trench off the west coast of South America, do not form back-arc basins. The back-arc basin that forms the Sea of Japan between China and the Japanese archipelago, which is a combination of ruptured continental fragments, will eventually be squeezed dry as the islands are plastered against Asia due to the interactions of lithospheric plates.

PLATE INTERACTIONS

When continental and oceanic plates converge, the denser oceanic plate dives underneath the lighter continental plate, forcing the oceanic plate farther downward. The sedimentary layers of both plates are squeezed like an accordion, swelling the leading edge of the continental crust to create folded mountain belts. The sediments are faulted at or near the surface, where the rocks are brittle, and folded at depth, where the rocks are ductile.

In the deepest parts of the continental crust, where temperatures and pressures are very high, rocks partially melt and become metamorphosed. As the descending plate dives farther under the continent, it reaches depths where the temperatures are extremely high. The upper part of the plate melts to form a silica-rich magma that rises toward the surface because of

Figure 9   The May 18, 1980, eruption of Mount St. Helens, Skamania County, Washington. Courtesy of USGS

its greater buoyancy. The magma intrudes the overlying metamorphic and sedimentary layers of the continental crust, either forming large granitic bodies or erupting on the surface.

Magma extruded onto the surface builds volcanic structures, including mountains and broad plateaus. The volcanoes of the Cascade Range came into existence when the Juan de Fuca plate subducted beneath the northwestern United States along the Cascadia subduction zone. As the plate melts while diving into the mantle, it feeds molten rock to magma chambers underlying volcanoes, such as Mount St. Helens, which produced one of the most powerful eruptions

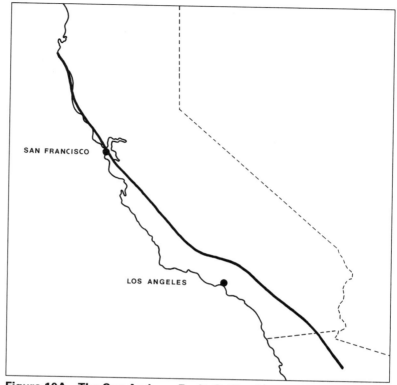

Figure 10A   The San Andreas Fault, California.

on the North American continent in this century (Fig. 9). The subducting plate also has the potential of generating very strong earthquakes, similar to those that occur in Chile and Japan.

When two continental plates collide, the topmost layers of the descending oceanic plate between them are scraped off and plastered against the swollen edge of the continental crust, forming an accretionary wedge. The submerged oceanic crust underthrusts the continental crust with additional crustal material, and the increased buoyancy raises mountain ranges like the Himalayas, which arose from the collision of the Indian and Eurasian plates. As a result, the Eurasian plate has shrunk some 1,000 miles since convergence began about 45 million years ago. Additional compression and deformation might develop farther inland beyond the line of collision, creating a high plateau with active volcanoes, like the wide Tibetan Plateau, which rises over 3 miles above sea level. The strain of raising the world's largest mountain range has resulted in strong deformation accompanied by powerful earthquakes all along the plate.

In submarine interactions, the divergence of lithospheric plates creates new oceanic crust, while convergence destroys old oceanic crust in subduction zones. Subduction of lithospheric plates is most prevalent in the western Pacific, where deep subduction zones are responsible for creating island arcs. Volcanoes of the island arcs are highly spectacular because their magmas are rich in silica and contain substantial amounts of volatiles, contrasting strongly with the highly fluid basalts of other volcanoes and spreading ridges. The volcanoes are highly explosive and build steep-sided cinder cones composed of a composite of cinder and lava. Island arcs are also associated with belts of deep-seated earthquakes several hundred miles below the Earth's surface.

Rifts open under continents as well as ocean basins. The best example is the great East African Rift Valley, which will eventually widen and flood with seawater to form a new subcontinent similar to Madagascar. This type of rifting is presently taking place in the Red Sea, where Africa and Saudi Arabia are diverging. The Gulf of Aden is a young oceanic rift between Africa and Arabia, which have been pulling away from each other for over 10 million years.

During the rifting process, large earthquakes strike the region as huge blocks of crust drop downward along diverging faults. Furthermore, volcanoes erupt due to the proximity of the mantle to the surface, which provides a ready supply of magma. A marked increase in volcanism produces vast quantities of basalt lava, which floods onto the continent during the early stages of rifting.

Sometimes an old extinct rift system, where the spreading activity has ceased, is overrun by a continent. For example, the western edge of the North American continent has overridden the northern part of the now extinct Pacific rift system, forming

Figure 10B   The San Andreas Fault in southern California. Photo by R. E. Wallace, courtesy of USGS

California's San Andreas Fault (Figs. 10A & 10B). A failed rift system beneath the central United States created the New Madrid Fault, which generated three tremendous earthquakes in the winter of 1811–1812. The potential of future great earthquakes in both of these regions remains dangerously high.

2

EARTHQUAKES

Earthquakes are by far the most destructive short-term natural force on Earth and have plagued civilizations for millennia. Powerful earthquakes destroy entire cities, often killing thousands of people with a single massive jolt (Table 1). The damage arising from a major earthquake is widespread, changing the landscape for up to thousands of square miles. Earthquakes often produce tall, steep-banked scarps and cause massive landslides that scar the countryside. Active faults crisscross much of the land surface at plate boundaries near the edges of continents. Half the world's population lives in coastal regions, where they are extremely vulnerable to earthquake destruction.

TABLE 1   THE 10 MOST LETHAL EARTHQUAKES SINCE 1900

Region (Date)	Death Toll
Tangshan, China (1976)	242,000
Kansu, China (1920)	180,000
Kwanto Plain, Japan (1923)	142,807
Messina, Italy (1908)	80,000
Kansu, China (1932)	70,000

Region (Date)	Death Toll
Yungay, Peru (1970)	66,800
Quetta, Pakistan (1935)	60,000
Northwest Iran (1990)	40,000
Erzincan, Turkey (1939)	30,000
Concepción, Chile (1935)	30,000

MAJOR QUAKES

The ancient Chinese were the first to keep accurate records on earthquakes because their country is so highly prone to devastating tremors. Perhaps the oldest recorded earthquake occurred in 1831 B.C. One of the most destructive earthquakes struck Shenshu in A.D. 1556, killing over 800,000

Figure 11   A crumpled bridge from the July 28, 1976, Tangshan, China, earthquake. Courtesy of USGS

people. The 1920 Kansu earthquake unleashed landslides that took the lives of some 180,000 people. The July 28, 1976, Tangshan earthquake left the city in ruins (Fig. 11) and killed an estimated 242,000 people. Unfortunately, it was not preceded by recognizable foreshocks that portend most great earthquakes, which would have provided ample warning and likely saved a great many lives.

Neighboring Japan, too, has been devastated by earthquakes throughout its history. Quakes in Tokyo took the lives of 200,000 people in 1802 and over 100,000 in 1857. On September 1, 1923, a magnitude 8.3 earthquake struck the Kwanto Plain in central Honshu. The wreckage was nearly total in Tokyo and Yokohama. Both cities were engulfed in massive firestorms that burned everything in their paths. When the disaster was over, 140,000 people had lost their lives. (Note that the earthquake magnitude scale is logarithmic; an increase of 1 magnitude signifies 10 times the ground motion and a release of 30 times the energy. Only a few earthquakes with a magnitude greater than 9 have been recorded.)

India has been rattled by earthquakes for ages, as well. On October 11, 1737, a strong earthquake hit Calcutta, killing about 300,000 people. One of the most powerful earthquakes in history struck the Assam region in northeast India on June 12, 1897. The area of total destruction covered about 9,000 square miles. A similar earthquake of magnitude 8.7 hit the region again on August 15, 1950, churning 10,000 square miles of landscape into desolation.

A magnitude 6.4 earthquake ripped through a large area of Maharashtra state in southwest India on September 30, 1993. It killed 12,000 people (some estimates are as high as 30,000), most of whom were asleep in stone and mud huts that crumbled down upon them, and left more than 100,000 homeless.

To the west, an earthquake of colossal destruction leveled Spitak, Armenia, and other neighboring cities in December 1988. The area lies atop an

Figure 12   Illustration of the collapsing city during the November 1, 1755, Lisbon, Portugal, earthquake. Courtesy of USGS

Figure 13 Damage to buildings viewed from the corner of Geary and Mason streets from the April 18, 1906, San Francisco, California, earthquake. Photo by W.C. Mendenhall, courtesy of USGS

active thrust fault created by tectonic forces deep in the bowels of the Earth. The poorly designed concrete buildings could not withstand the earthquake's terrible punishment as the ground suddenly lifted as much as 6 feet. At least 25,000 people lost their lives when buildings crashed down on them. Nearby, in northern Iran, a powerful earthquake in June 1990 triggered massive landslides that killed at least 40,000 people and left half a million homeless.

Continuing westward, a tremendous force was unleashed against Lisbon, Portugal, on November 1, 1755, causing buildings to crumble and fall to the ground (Fig. 12). A vast inferno that burned for several days enveloped the city, turning it to ashes. In addition, a 20-foot tsunami generated by the undersea quake swept through the harbor, overturning ships. The earthquake completely leveled the city and killed 60,000 of its inhabitants. The powerful quake might have triggered sympathetic tremors in North Africa, causing heavy damage there as well.

Three of the greatest earthquakes to strike the continental United States shook New Madrid, Missouri, on December 16, 1811, January 23, 1812, and

February 7, 1812. The town was the largest settlement in the region, and the quakes thoroughly destroyed nearly all buildings. Trees snapped in two, the ground split open forming deep fissures, and massive landslides marred the countryside. The earthquakes even changed the path of the Mississippi River, which wandered far to the west of its normal course, while the downdropped crust formed deep lakes. If similar quakes hit the same area today, the damage would be colossal because the region now contains many large cities, with a total population of about 12 million.

The largest earthquake in California's recent history ruptured the southern end of the San Andreas Fault near Los Angeles in 1857. Nearly 50 years later, on April 18, 1906, the northern end of the fault ruptured near San Francisco, causing considerable damage to the city (Fig. 13) and taking as many as 3,000 lives. The town of Santa Rosa, 50 miles to the north, was almost totally wrecked by the magnitude 7.9 earthquake. A repeat performance played out at San Francisco on October 17, 1989, although the damage and the loss of life was not nearly as great. Had this earthquake been as powerful as the one in 1903, damages could have reached $40 billion, with tremendous casualties.

On Good Friday, March 27, 1964, the largest recorded earthquake to hit the North American continent devastated Anchorage, Alaska and surrounding areas. The magnitude 9.2 quake caused destruction over an area of 50,000 square miles and was felt over an area of half a million square miles. Landslides caused much of the damage, and entire port facilities slid bodily into the

Figure 14 The Fourth Avenue slide area in Anchorage from the March 27, 1964, Alaskan earthquake. Courtesy of USGS

sea. Anchorage bore the brunt of the earthquake, and 30 blocks were destroyed when the city subsided several feet (Fig. 14). A 30-foot-high tsunami generated by the undersea earthquake destroyed coastal villages around the Gulf of Alaska killing 107 people. Kodiak Island was heavily damaged, and most of the fishing fleet was destroyed when the tsunami carried many vessels inland.

Latin America has had more than its fair share of destructive earthquakes. The May 22, 1960, Chilean earthquake of magnitude 9.5 was the most powerful ever recorded. It unleashed a wave of destruction over an area of 90,000 square miles, destroying some 50,000 homes and taking nearly 6,000 lives. Three other earthquakes of similar size have shaken the area during the last 400 years. More recently, the December 23, 1972, Managua, Nicaragua, earthquake killed some 10,000 people. The 1976 Guatemala City earthquake took the lives of about 23,000 people. And on September 19, 1985, an earthquake called the gravest disaster in Mexico's history toppled buildings in Mexico City and killed upwards of 10,000 people.

AREAS AT RISK

Most earthquakes originate at plate boundaries (Fig. 15), where lithospheric plates are converging, diverging, or sliding past each other. The most powerful earthquakes are associated with plate subduction, when one plate thrusts under another in deep subduction zones. Oceanic and continental rifts, like the East African Rift, also induce strong earthquakes. The greatest amount of seismic energy is released along the rim of the Pacific Ocean, known as the circum-Pacific belt, which is a band of subduction zones that flank the Pacific Basin. It coincides with the Ring of Fire, and explains why the Pacific rim also contains most of the world's active volcanoes.

In the western Pacific, the circum-Pacific belt encompasses the volcanic island arcs that fringe sub-

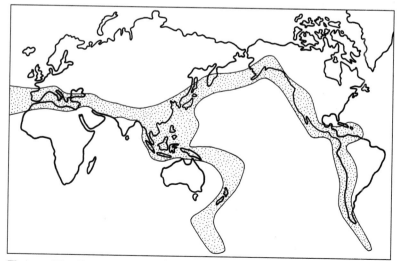

Figure 15 Earthquake belts associated with plate boundaries, where most seismic activity occurs.

duction zones that produce some of the largest earthquakes on Earth. The belt continues along the Aleutian Trench, responsible for many great Alaskan earthquakes, the Cascadia subduction zone, which has shaken the Pacific Northwest, and the San Andreas Fault that rattles southern California.

Continuing in the eastern Pacific, the belt runs down the Andes Mountains of Central and South America, known for many of the largest and most destructive earthquakes on record. During this century, nearly two dozen earthquakes with magnitudes of 7.5 or greater have devastated these areas. An immense subduction zone just off the coast threatens the entire western seaboard of South America. The lithospheric plate on which the South American continent rides is forcing the Nazca plate to buckle under, building up great tensions deep within the crust. While some rocks are forced downward, others are pushed to the surface to raise the Andes Mountains, the fastest-growing mountain range on Earth. The resulting forces are building great stresses into the entire region. When the strain becomes too great, earthquakes roll across the countryside.

A second major seismic zone runs through the folded mountain regions that flank the Mediterranean Sea. The collision of Africa and Eurasia upraised the mountains during the Alpine orogeny (mountain building episode) 26 million years ago. The belt continues through Iran and past the Himalayan Range into China. At the eastern end of the Himalayas lies possibly the most earthquake-prone area in the world. An enormous 2,500-mile-long seismic zone stretches across Tibet and much of China. In this century, more than a dozen earthquakes with magnitudes of 8.0 or greater have struck the region.

Westward, the Hindu Kush Range of northern Afghanistan is the site of many devastating earthquakes. During this century, the region has witnessed three great earthquakes of magnitude 8.0 or greater. This is a highly active earthquake belt, with some 2,000 minor quakes occurring annually. The seismic zone continues through the Caucasus Mountains on to Turkey. The eastern Mediterranean region is a jumble of colliding plates, providing highly unstable ground. The Near East is extremely unstable, attested by many earthquakes reported in biblical times. Earthquakes have often ravaged the remaining regions surrounding the Mediterranean.

Earthquakes also occur in stable zones that comprise the strong rocks in the interiors of the continents (Fig. 16). The stable zones include Scandinavia, Greenland, eastern Canada, parts of northwestern Siberia and Russia, Arabia, the lower portions of the Indian subcontinent, the Indochina Peninsula, almost all of South America except the Andean mountain region, much of Australia, and the whole of Africa except the Great Rift Valley and northwestern Africa. Earthquakes in these regions might have been triggered by the weakening of the crust by compressive forces originating at plate edges. The crust might have been weakened by previous

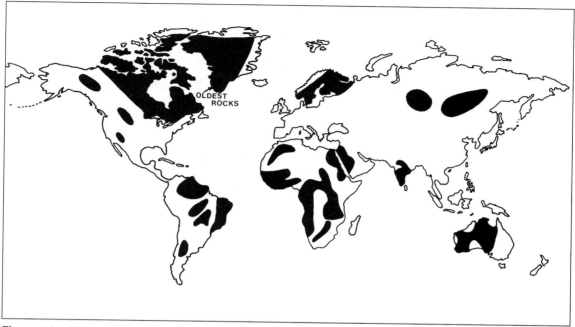

Figure 16   Areas of exposed Precambrian rocks in the continental interiors.

tectonic activity, involving extinct or failed rift systems similar to the New Madrid Fault, which, according to geologic history, is poised for another major quake.

EARTHQUAKE FAULTS

The mechanics of earthquakes were not well understood until after the great 1906 San Francisco quake that occurred along the southern San Andreas Fault. For hundreds of miles, roads and fences crossing the fault were offset by as much as 21 feet (Fig. 17). The San Andreas Fault runs 650 miles from the Mexican border through the western edge of California and plunges into the Pacific Ocean near Cape Mendocino in the northern part of the state. It represents the boundary between the Pacific and North American plates. The segment of California west of the San Andreas Fault is sliding past the North American continent in a northwestward direction about 2 inches per year. The relative movement of the two plates is right lateral (Fig. 18) or dextral because an observer on either side of the fault would notice the other block moving to the right. During the San Francisco earthquake, the Pacific plate suddenly slipped several feet relative to the North American plate.

During the 50 years since the last major earthquake struck in 1857, the rocks along the southern San Andreas Fault were bending and storing elastic energy like a stick stores elastic energy when it is bent. Eventually, the forces holding the rocks together lost their strength and slippage occurred at the weakest point. The displacement exerted strain further along the fault until most of the built up strain was released, and like a stick bent to its maximum, it snapped.

Such sudden slippage allows deformed rock to rebound. As the rock elastically returns to its original shape, it produces vibrations called seismic waves, which are similar to sound waves. The seismic waves radiate in all directions, like those generated by tossing a pebble into a quiet pond. The rocks do not always rebound immediately, however, and could take days or even years, resulting in slippage not associated with earthquakes. This condition generally occurs along the midsection of the San Andreas Fault, whereas at the southern end of the fault, responsible for the massive 1857 quake, and at the northern end of the fault, known as the "big bend," earthquakes rumble across the landscape when locked portions of the fault attempt to tear free.

Figure 17   A fence in Marin County offset 8.5 feet by the San Andreas Fault during the April 18, 1906, San Francisco, California, earthquake.
Photo by G. K. Gilbert, courtesy of USGS

An unusual set of circumstances caused by lateral slippage at the southern end of the San Andreas Fault combined with plate subduction at the northern end resulted in a major earthquake of magnitude 6.1 in southern California near Joshua Tree on April 22, 1992, and two more to the north three days later. The largest quake to hit California in 30 years was the June 28 Landers earthquake of magnitude 7.5. Three days later, a magnitude 6.5 earthquake shook the town of Big Bear, 20 miles to the northwest. Luckily, the strongest seismic waves raced farther northward into the sparsely populated Mojave Desert rather than heading into the crowded urban areas to the west.

The tremors might herald a great quake, the "Big One," as Californians say. They could have had a direct effect on a nearby segment of the San Andreas Fault by shifting stresses in the crust, which make conditions easier for an earthquake to occur along the fault. The quakes also might signal the birth of a new competing fault, parallel to the San Andreas. The midsection of the San Andreas Fault produces relatively mild earthquakes because the two sides of the fault generally slide smoothly past each other. Conversely, the ends of the fault tend to snag, and any slippage generates powerful earthquakes.

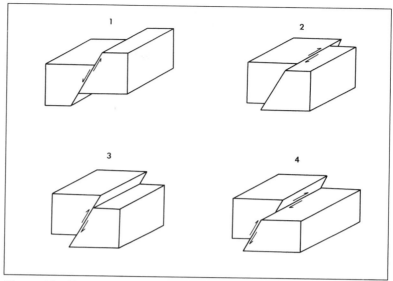

Figure 18   Fault types: 1. normal fault, 2. lateral fault, 3. thrust fault, 4. oblique fault.

Faults are also displaced in a vertical direction, with one side positioned higher with respect to the other side (see Fig. 18). When the crust is pulled apart, a fault block slides downward along an inclined plane, resulting in a gravity or normal fault. However, most faults arise from compressional forces that push fault blocks upward along an inclined plane. These are called reverse faults because they are the opposite of gravity faults. If the reverse fault plane is nearly flat and the movement is mostly horizontal, it produces a thrust fault. A thrust fault forms when a compressed plate shears, causing one section to be lifted over another for large distances.

Thrust faults are not always exposed on the surface. Some thrust faults associated with the San Andreas Fault lie deep underground, where stresses along the fault increase with depth. A thrust fault lying about 6 miles beneath the surface produced a magnitude 6.7 earthquake at Coalinga, California on May 2, 1983, that nearly leveled the town. The unusual earthquake had no recognizable foreshocks that would have provided advance warning. The January 17, 1994, Northridge, California, earthquake of magnitude 6.6 occurred on a thrust fault that caused severe damage in the Los Angeles area, killing 70 people and leaving 10,000 homeless. Thrust faults associated with the San Andreas Fault system might be expressed on the surface as a series of active folds that continuously uplift California's Coast Ranges.

Thrust faults can cause considerably more damage than lateral faults with equal measures of magnitude. Lateral faults cause buildings to sway back and forth, allowing their flexible frames to absorb most of the force. Some buildings sway so fiercely during an earthquake they seriously injure occupants, as furniture and other objects (including people) are hurled against walls. Thrust faults suddenly raise and drop buildings inches or feet at a time, creating tremendous forces that topple even the best-designed structures. A thrust fault responsible for the 1988 Armenian earthquake caused substantial damage and the loss of a great many lives.

Some faults are a combination of horizontal and vertical motions, consisting of complicated diagonal movements that form complex fault systems called oblique or scissors faults. The 1989 Loma Prieta, California, earthquake (Fig. 19) ruptured a 25-mile-long segment of the San Andreas Fault. The faulting propagated upward along a dipping plane, resulting in a right oblique reverse fault. The earthquake raised the southwest side of the fault over 3 feet, contributing to the continued growth of the nearby Santa Cruz Mountains.

Figure 19   Buildings damaged in the Marina district, San Francisco, from the October 17, 1989, Loma Prieta, California, earthquake. Photo by G. Plafker, courtesy of USGS

Both the 1906 San Francisco and the Loma Prieta earthquakes took place on a segment of the San Andreas Fault that runs through the Santa Cruz Mountains. The major difference between the two is that most of the motion of the earlier quake was horizontal, whereas the latter quake occurred along a tilted surface that forced the southwest side of the fault to ride over the northeast side. Furthermore, because of the subsurface geology of the region, the shaking of the San Francisco area reached double the level expected for an earthquake of this size.

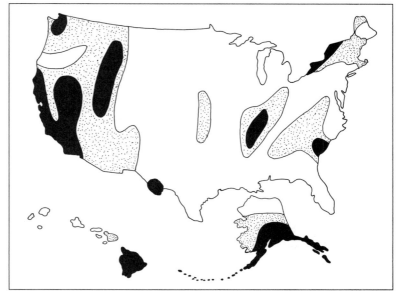

Figure 20   Earthquake risk areas. Shaded areas can expect major damage. Stippled areas can expect moderate earthquakes.

The rest of the nation is crisscrossed by numerous faults generally associated with mountains, and 39 states lie in regions classified as having moderate to major earthquake risk (Fig. 20). The Basin and Range Province in Oregon, Nevada, western Utah, southeastern California, and southern Arizona and New Mexico contains several fault block mountains that are prone to earthquakes. The upper Mississippi and Ohio river valleys suffer frequent earthquakes. The northeast trending New Madrid and associated faults are responsible for major earthquakes and many tremors. Along the eastern seaboard, major earthquakes have hit Boston, New York, Charleston, and other areas. Since colonial times, eastern North America has had over a dozen moderate to large earthquakes. This century has been comparably quiet in this region, spurring fears that a major earthquake might be looming just over the horizon.

EARTHQUAKE CAUSES

The energy released by a moderate earthquake is equivalent to a hundred Hiroshima-sized atomic bombs. The damage caused by earthquakes is widespread, covering thousands of square miles. Besides the destruction of man-made structures, earthquakes alter the landscape by producing deep fissures and tall scarps (Fig. 21) and by causing massive landslides that scar the terrain. Successive earthquakes also can upraise large folds. For example, an anticline (upfolded strata) associated with the fault responsible for the 1980 El Asnam, Algeria, earthquake was uplifted over 15 feet.

Earthquakes associated with folds do not rupture the Earth's surface the way faults do, though many of the world's major fold belts that raised mountain ranges, like those bordering the Mediterranean Sea, are earthquake prone. During this century, large fold earthquakes have taken place in Japan, Argentina, New Zealand, Iran, and Pakistan. Most of these earthquakes appear to have occurred under young anticlines less than a few million years old. This is because folds are the geologic product of successive earthquakes arising from compressional forces during plate collisions.

Why seismic energy is released violently in some cases and not in others is not fully understood. Generally, the deeper and longer the breaking fault the larger the earthquake. The great 1960 Chilean earthquake, the largest in this century, was generated along a 600-mile-long rupture through the South Chilean subduction zone, which might have slipped all at once. During the 1906 San Francisco earthquake, a 260-mile section of the San Andreas Fault ruptured. Other processes that affect the magnitude of an earthquake include the speed of the rupture as it travels over the fault (a break along a fault can move at speeds of up to a mile a second), the frictional strength of the fault, and the drop in stress across the fault.

The Earth's crust is constantly readjusting itself, producing vertical and

Figure 21   The Red Canyon Fault scarp from the August 1959 Hebgen Lake earthquake is about 14 feet high, Gallatin County, Montana.
Photo by I. J. Witkind, courtesy of USGS

Figure 22   Wreckage from the August 31, 1886, Charleston, South Carolina, earthquake. Photo by J. K. Hillers, courtesy of USGS

horizontal offsets on the surface. These movements are associated with large fracture zones in the crust. The greatest earthquakes have offsets of several tens of feet occurring in a matter of seconds. Most faults are associated with plate boundaries, and most earthquakes are generated in zones where huge plates collide or shear past each other. The interaction of plates causes rocks to strain and deform. If deformation takes place near the surface, major earthquakes result. Earthquakes also occur during volcanic eruptions, but they are relatively mild compared to those caused by faulting.

Thousands of earthquakes strike yearly. Fortunately, only a few are powerful enough to be destructive. During this century, the world average was about 18 major earthquakes of magnitude 7.0 or greater per year. For great earthquakes with magnitudes above 8.0, the century's average is 10 per decade. The degree of damage is not dependent on magnitude alone

Figure 23   Collapse of the Terminal Hotel, caused by the failure of reinforced concrete columns during the 1976 Guatemala City, Guatemala, earthquake. Courtesy of USGS

but is also influenced by the geology of the region. Earthquakes in strong rocks, such as those in continental interiors, are more destructive for equal magnitudes than those occurring in the fractured rock at plate margins. This is why earthquakes in the eastern United States influence a wider area than earthquakes in the West. The August 31, 1886, Charleston, South Carolina, earthquake (Fig. 22) that killed 110 people cracked walls in Chicago 750 miles away and was felt in Boston, Milwaukee, and New Orleans.

The longer the time since the last big shock on a major fault the greater the earthquake hazard. This is known as the seismic gap hypothesis, which holds that the earthquake hazard along major faults is low immediately following a large earthquake and increases with time. The earthquake hazard varies depending on the region and the size of previous tremors in the region. Most faults appear to have a characteristic earthquake that recurs in a similar form. Some areas might experience earthquakes of magnitude 7, whereas other regions might be prone to great earthquakes of magnitude 8 or 9. However, larger quakes do not follow the same patterns set by smaller ones, making their prediction extremely difficult. Once a zone becomes seismically active, earthquakes continue until, for unknown reasons, they stop. Then a relatively long interval passes before another great one occurs.

EARTHQUAKE DAMAGE

Ancient civilizations living in earthquake-prone regions protected themselves from the ravages of quakes by constructing simple dwellings that could withstand violent shaking. Today, however, as accommodations have become more sophisticated with complex construction materials, earthquake damage has become a serious and expensive problem (Fig. 23). Most large urban centers are a combination of old and new buildings, often with modern structures blending in with turn-of-the-century architecture, whose foundations have weakened with time.

Many areas like the Pacific Northwest, which lies along a subduction zone and has been devastated by earthquakes in the distant past, were not built to survive severe ground shaking and are totally unprepared for a major quake. About 1,000 years ago a major earthquake struck the Seattle, Washington, area, and the ground shook with such fury that avalanches and landslides tumbled from the Olympic Mountains and buried areas that are now densely populated. The earthquake also triggered a great tsunami that washed the shores of the Puget Sound.

Overcrowding and high real estate prices have forced builders to construct towering blocks of masonry that dominate the skyline. To save money, designs and materials might not always conform to building codes in earthquake-prone areas. The type of construction determines how well a structure survives an earthquake. Lightweight, steel-framed buildings with strength and flexibility, and reinforced concrete buildings with few structure-weakening window and door openings generally suffer little damage during an earthquake.

A building's ability to withstand a major earthquake depends not only on its design and type of materials used, but also its orientation with respect to the shock wave and the nature of the shock wave. A short, sharp high-frequency shock wave lasting only a few seconds is comparatively easy to design for. Buildings of two to four stories are most vulnerable to this type of shock wave, whereas taller buildings might escape unscathed. A longer, lower frequency shock wave lasting for several tens of seconds is much more difficult to design for. Multistory buildings are most vulnerable to this type of shock wave, whereas shorter buildings are practically untouched.

Even if buildings are able to withstand an earthquake, they are still vulnerable to foundation failure, which causes buildings to topple over when the ground gives way beneath them. Severe shock waves can make soils settle or liquefy, causing them to lose their ability to support structures (Fig. 24). The building site greatly affects the amount of movement a structure experiences. Generally, structures built

Figure 24   Highway 1 bridge destroyed by liquefaction of river deposits at Struve Slough during the October 17, 1989, Loma Prieta, California, earthquake. Photo by G. Plafker, courtesy of USGS

on bedrock are damaged less severely than those built on less consolidated, easily deformed material such as natural and artificial fills. The type of ground that supports a structure affects the amount of movement because soft sediments generally absorb high-frequency vibrations and amplify low-frequency vibrations, which do the most damage.

The length of time an earthquake is in motion greatly affects the amount of damage to buildings, and generally the longer the earthquake interval the more severe the damage. Other factors that determine the degree of earthquake destruction include the type of seismic waves involved. As the energy released by an earthquake travels along the surface, it causes the ground to vibrate in a complex manner, moving up and down as well as from side to side. Most buildings can handle the vertical motions because they are built against the force of gravity. However, the largest ground motions are usually horizontal, which causes buildings to sway back and forth. If the structure's resonance frequency is equal to that of the earthquake, it could sway wildly and cause considerable damage. Aftershocks caused by readjustments in rocks following the main event can be just as destructive and finish off what the earthquake started. Further damage is caused by fires set in broken gas lines and other flammable materials that burn out of control, with fire fighting efforts hampered by broken water mains.

The size of the geographic area influenced by shock waves depends on the magnitude of the earthquake and the rate at which the amplitudes of the seismic waves diminish with distance. Some types of ground transmit seismic energy more effectively than others. For a given magnitude, seismic waves extend over a much wider area in the eastern United States than in the West (Fig. 25), which indicates a substantial difference in the crustal composition and structure of the two regions. The East is composed of older sedimentary rock, while the West is composed of fairly young igneous and sedimentary rocks that are fractured by faults.

In the future, a system involving a network of sensors that detect seismic waves before they spread from the epicenter of an earthquake could provide up to 60 seconds of advance warning to outlying

Figure 25   Comparison between western and eastern earthquake area destruction.

areas. A central computer would process data transmitted by seismic sensors to determine the size and location of the quake and send out information to areas in the path of damaging vibrations. The early warning would activate automated systems, for example shutting off gas lines to prevent fires, that could save lives and property. The system also would provide information to help emergency officials locate the sites hardest hit by the earthquake.

SEISMIC SEA WAVES

Undersea earthquakes that vertically displace the ocean floor produce seismic sea waves, called tsunamis, the Japanese word for "tidal waves"; they are so named because of their common occurrence in that country.

Figure 26   Tsunamis washed many vessels into the heart of Kodiak from the March 27, 1964, Alaskan earthquake.   Courtesy of USGS

The energy of an undersea earthquake transforms into wave energy proportional to its intensity. The earthquake sets up ripples on the ocean like those formed by tossing a rock into water. In the open ocean, the wave crests are up to 300 miles long and usually less than 3 feet high, with a distance between crests of 60 to 120 miles. This gives the tsunamis very gentle slopes that pass practically unnoticed by ships or aircraft.

A tsunami travels at a speed of 300 to 600 miles per hour. When it touches bottom upon entering shallow coastal waters, such as a harbor or narrow inlet, its speed diminishes rapidly to about 100 miles per hour. The sudden breaking action causes seawater to pile up and the wave height is magnified tremendously. Tsunamis have been known to grow into a towering wall of water up to 200 feet high, although most are only a few tens of feet high. The destructive power of the wave is immense, causing considerable damage as it crashes to shore. Buildings are crushed with ease and ships are often carried well inland (Fig. 26).

Prior to the establishment of a tsunami watch in the Pacific Ocean, which is responsible for 90 percent of all tsunamis in the world, people had little warning of impending disaster except for a rapid withdrawal of seawater from the shore. Residents of coastal areas frequently stricken by tsunamis have learned to heed this warning and head for higher ground. When a tsunami struck on the island of Madeira during the 1755 Lisbon, Portugal, earthquake, large quantities of fish were stranded on shore as the sea suddenly retreated. Villagers, unaware of any danger, went out to collect this unexpected bounty, only to lose their lives when without warning a gigantic wave crashed down on them.

A few minutes after the sea retreats, a tremendous surge of water pounds the shore, extending hundreds of feet inland. Often a succession of surges occurs, each followed by a rapid retreat of water back to sea. On coasts and islands where the seafloor rises gradually or where barrier islands exist, much of the tsunami's energy is spent before it ever reaches shore. However, on volcanic islands surrounded by very deep water or where deep submarine trenches lie immediately outside harbors, an oncoming tsunami can build to tremendous heights.

Destructive tsunamis generated by large earthquakes can travel clear across the Pacific Ocean. The great 1960 Chilean earthquake created a 35-foot-high tsunami that struck Hilo, Hawaii, over 5,000 miles away, causing more than $20 million in property damages and 61 deaths. The tsunami traveled an additional 5,000 miles to Japan and inflicted considerable destruction on the coastal villages of Honshu and Okinawa, leaving 180 people dead or missing. In the Philippines, 20 people were killed. Coastal areas of New Zealand were also damaged. For several days afterward, tidal gauges in Hilo could still detect the waves as they bounced around the Pacific Basin.

3

VOLCANIC ERUPTIONS

Volcanoes are the second most destructive natural force on Earth and constitute a major geologic hazard (Table 2). Two-thirds of all historic eruptions have caused fatalities. Since the ice age ended around 12,000 years ago, some 1,300 volcanoes are known to have erupted. During the past 400 years, over 500 volcanoes have erupted, killing more than 200,000 people and causing billions of dollars in property damage. Since the year 1700, 23 volcanoes have earned special recognition for killing more than 1,000 people each. During the 20th century, volcanoes have caused an average death toll of over 800 people per year. In the 1980s alone, some 30,000 people have lost their lives to volcanic eruptions.

TABLE 2   GREATEST VOLCANIC ERUPTIONS: TNT EQUIVALENT AND MATTER EJECTED

Volcano (Date), Region	Megatonnage (1 megaton [Mt] = 1 million tons of TNT)	Matter Ejected (km^3)
Tambora (1815), Sumbawa, Indonesia	20,000 Mt	40 (excluding collapsed material)

Volcano (Date), Region	Megatonnage (1 megaton [Mt] = 1 million tons of TNT)	Matter Ejected (km^3)
Thera (1628? B.C.), Cyclades, Greece	7,500 Mt	50–65
Krakatoa (1883), Java, Indonesia	1,500 Mt	20
(Atomic bomb dropped on Hiroshima, Japan)	0.02 Mt	

The increase in volcano-related deaths is mainly due to swelling populations living near active volcanoes and not necessarily more eruptions. The majority of deaths occur in developing countries, which often do not have the technology to provide advance warning. Presently, about 600 active volcanoes—meaning they have erupted in historic times—exist throughout the world, with many thousand more dormant or extinct ones. However, even a volcano that has been dormant for up to a million years or more is not totally immune to eruption.

VIOLENT VOLCANOES

What could well be the most explosive volcanic eruption in recorded history took place in the early to middle 1600s B.C. on the island of Thera, 75 miles north of Crete in the Mediterranean Sea. The eruption might have caused the demise of the Minoan civilization on Crete and surrounding islands. A huge magma chamber beneath the island apparently flooded with seawater, and like a gigantic pressure cooker the volcano blew its lid. The island then collapsed into the emptied magma chamber, forming a deep, gaping caldera that covered an area of about 30 square miles (Fig. 27). The collapse of Thera created a colossal tsunami that battered the shores of the eastern Mediterranean, no doubt causing additional death and destruction.

The cataclysm was so violent it could have been heard as far away as Scandinavia. An immense ash cloud drifted as far south as Egypt and might have been responsible for the biblical Egyptian plagues. In the meantime, according to the book of Exodus, when Moses led the Israelites out of Egypt, "The Lord guided them by a pillar of cloud during the daytime and by a pillar of fire at night. The cloud and fire were never out of sight." Some historians believe the Greek poet Homer based some of his legends on the disaster, as did the Greek philosopher Plato, with his story about the lost island of Atlantis.

On August 24, A.D. 79, the seaward side of Mount Vesuvius, Italy, blew outward, sending searing hot ash mixed with steam and deadly gases toward the city of Pompeii. The volcanic debris buried people almost immediately, and as many as 16,000 suffocated to death. At the same time Pompeii was being buried under 20 feet of ash, the nearby town of Herculaneum was submerged in a sea of boiling mud. Another violent eruption of Vesuvius in 1631 destroyed several nearby villages and took 18,000 lives. The volcano struck again in 1661, taking the lives of 4,000 people in Naples. In 1740, the

Figure 27 The caldera formed by the eruption of Thera.

buried city of Pompeii was rediscovered by a farmer digging a water well. When archaeologists began excavation, they found a gruesome reminder of the volcano's fury, expressed by the stark terror on the faces of fossilized inhabitants.

About 200 miles to the south on the island of Sicily lies Mount Etna, the largest volcano in Europe. Records of its eruptions kept by the early Greeks go back several centuries B.C. During its long history, Etna's numerous outbursts have destroyed many towns and taken thousands of lives. The 1169 eruptions buried 15,000 people in the ruins of the nearby town of Catania, and 500 years later in 1669, 20,000 people perished. Etna struck again in 1853, and in 1928 it demolished the towns of Mascati and Nunziata. Yet the townspeople always returned, refusing to give up their farms on the rich volcanic soil. Mount Etna again showed its fury in April 1992, when lava flows threatened the Sicilian town of Zafferana Etnea, whose townspeople battled desperately against the fury of the volcano and turned the lava away.

Indonesian volcanoes are among the most explosive in the world and have produced more violent blasts in historic times than those of any other region (Fig. 28). On April 11, 1815, Tambora on the island of Sumbawa created one of the most explosive volcanic eruptions in the last 10,000 years, spewing some 25 cubic miles of debris into the atmosphere. It cast more volcanic dust into the air and obscured more sunlight than any other volcano in the past 400 years. The volcano took the lives of nearly 60,000 people locally and caused climatic havoc, starvation, and disease around the rest of the world. The lowered temperatures and killing frosts during the summertime resulted in "the year without summer."

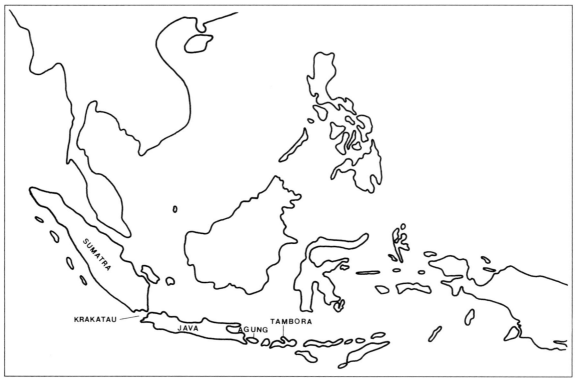

Figure 28   Locations of the great Indonesian volcanoes.

The volcanic island of Krakatau, located between Java and Sumatra, was nearly decimated on August 27, 1883, by a series of tremendous explosions. The eruptions might have been powered by a rapid expansion of steam from seawater entering a breach in the magma chamber. After the last convulsion, most of the island collapsed, creating a large undersea caldera over 1,000 feet below sea level. The sound of the explosion carried as far as Madagascar, 3,000 miles away—it was called the loudest noise known to mankind. Barographs around the world recorded the atmospheric pressure wave as it circled the Earth at least three times. In the nearby coastal areas, the eruption produced towering tsunamis over 100 feet high that swept 36,000 people to their deaths.

Mount Pelée, located on the island of Martinique in the West Indies, violently erupted on May 8, 1902. The explosion ripped out the seaward side of the volcano, and a solid sheet of flame rolled down the mountain and headed for the port city of St. Pierre. Clouds of hot ashes and suffocating fumes swept through the city and out to sea. The hot blast set fire to everything it touched (Figs. 29A & 29B), even ships in the harbor. Practically the entire population of 28,000 died in less than 3 minutes. Nearly all the victims were found with their hands covering their mouths or in some other agonizing posture, showing they had perished by suffocation. That same year, 15,000 people lost their lives to the eruption of La Soufrière on the nearby island of St. Vincent.

The violent explosion of Mount St. Helens, Washington, on May 18, 1980, was the largest volcanic eruption in the continental United States in several centuries (Fig. 30).

Figures 29A & 29B   Devastation at St. Pierre, Martinique, West Indies, from the May 8, 1902, eruption of Mount Pelée.   Photo by I. C. Russell, courtesy USGS

Mount St. Helens lies in the Cascade Range, stretching from northern California to southern British Columbia, along with 15 other major active volcanoes. The explosion blew off the top one-third of the mountain and lofted a cubic mile of debris into the atmosphere. The volcano produced perhaps the largest landslides in recorded history and created a massive mudflow and flood that reached all the way to the Pacific Ocean. The blast devastated the nearby forest, toppling trees like scattered matchsticks. The

Figure 30   The explosive eruption of Mount St. Helens, Washington, on May 18, 1980. Courtesy of USDA

eruption also marked the beginning of a decade of increased volcanic activity throughout the world.

The March 28, 1982, eruption of El Chichon in Chiapas, southern Mexico (Fig. 31), is counted among the dirtiest volcanoes in recent history. The volcano killed 187 people and left 60,000 homeless. El Chichon blasted a gigantic ash cloud high into the stratosphere, and several inches of ash covered nearby regions. In three weeks, the ash cloud encircled the entire globe and lowered global temperatures by as much as 0.5 degree Celsius.

Figure 31   Caldera formed by the March 28, 1982, massive eruption of the El Chichon volcano, Chiapas, Mexico. Courtesy of USGS

A similar type of eruption occurred at Mount Pinatubo, Philippines, on June 15, 1991, possibly the largest outburst of the century. The eruption spewed up to 20 million tons of weather-altering sulfur dioxide 22 miles into the atmosphere, about twice as much as El Chichon. Some 700 people died during the first three months of eruptions, and tens of thousands of families lost their homes. Deep ash buried two important American military bases near the volcano that had to be permanently abandoned. The eruption dropped global temperatures possibly as much as 0.5 degree Celsius and was blamed for the strange weather that occurred worldwide in 1992, referred to as the "Pinatubo winter." Furthermore, the sulfuric acid aerosol from Pinatubo might have aided in the depletion of the upper-atmospheric ozone layer, which shields the Earth from harmful solar ultraviolet radiation.

The November 13, 1985, eruption of the Nevado del Ruiz Volcano in Colombia, the second worst volcanic disaster of the century, created one of the world's largest volcanic mudflows. The eruption melted the mountain's icecap and sent floods and mudflows cascading 90 miles per hour down the volcano's sides. A 130-foot-high wall of mud and ash careened down the narrow canyon below and headed for the town of Armero 30 miles away. When the mudflow reached the town, 10-foot-high waves spread out and flowed rapidly through the streets. The mass of mud buried almost all the town along with nearby villages, killing more than 25,000 people.

THE FIRE BELOW

The interaction of lithospheric plates on the Earth's surface is responsible for generating most of the world's volcanic activity. Subduction zones created by descending plates accumulate large quantities of sediment from the adjacent continents and islands. The sediments are carried deep into the mantle, where they melt in pockets called diapirs. The diapirs rise toward the surface to form magma bodies, which become the source for new igneous activity (Fig. 32).

The magma also might have originated from the partial melting of subducted oceanic crust, with heat supplied by the shearing action at the top of the descending plate. Convective motions in the wedge of asthenosphere caught between the descending oceanic plate and the continental plate force material upward, where it melts under lowered pressures.

The mantle material that extrudes onto the surface is black basalt, and most of the 600 active volcanoes in the world are entirely or predominately basaltic. The mantle material below spreading ridges, which create new oceanic crust, consists mostly of peridotite, which is rich in silicates of iron and magnesium. As the peridotite melts on its way to the surface, a portion becomes highly fluid basalt. The magma that forms basalt originates in a zone of partial melting in the upper mantle more than 60 miles below the surface. The semimolten rock at this depth is less dense and therefore more buoyant than the surrounding mantle material and rises slowly toward the surface. As the magma ascends, the pressure decreases and more mantle material melts. Volatiles, such as dissolved water and gases, make the magma flow easily.

As the magma rises toward the surface, it replenishes shallow reservoirs or feeder pipes that are the immediate sources for volcanic activity. The magma chambers closest to the surface exist under spreading ridges where the oceanic crust is only 6 miles or less thick. Large magma chambers lie under fast spreading ridges that cre-

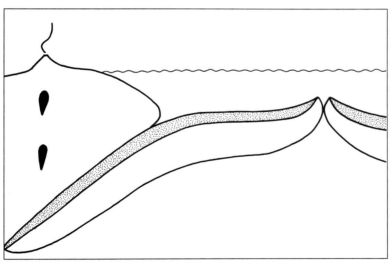

Figure 32   The subduction of a lithospheric plate into the mantle furnishes volcanoes with molten magma, which rises to the surface in blobs called diapirs.

ate new lithosphere at a high rate, such as those in the Pacific. Conversely, narrow magma chambers lie under slow spreading ridges, such as those in the Atlantic.

When the magma chamber swells with magma and begins to expand, it pushes the crest of the spreading ridge upward by the buoyant forces generated by the molten rock. The magma rises in narrow plumes that mushroom out along the spreading ridge, welling up as a passive response to plate divergence. Only the center of the plume is hot enough to rise all the way to the surface. If the entire plume were to erupt, it could build a massive volcano several miles high that would rival the tallest volcanoes in the Solar System.

The composition of the magma is indicative of its source materials and the depth within the mantle from which they originated. Degrees of partial melting of mantle rocks, partial crystallization that enriches the melt with silica, and assimilation of a variety of crustal rocks in the mantle influence the composition of the magma. When the erupting magma rises toward the surface, it incorporates a variety of rock types along the way. This changes the magma's composition, which is the major controlling factor in determining the type of eruption.

The composition of the magma also determines its viscosity and whether it erupts mildly or explosively. If the magma is highly fluid and contains little dissolved gas when it reaches the surface, it produces basaltic lava, and the eruption is usually quite mild. The two types of lava from this type of eruption are aa, or blocky, lava and pahoehoe, or ropy lava, which are Hawaiian names. However, if the magma rising toward the surface contains a large quantity of dissolved gases, it erupts in a highly explosive manner that can be quite destructive.

VOLCANIC ACTIVITY

The majority of volcanoes are associated with crustal movements at the margins of lithospheric plates. When one plate subducts under another, the lighter rock component melts and rises toward the surface in giant blobs of magma. The molten rock subsequently feeds magma chambers lying below active volcanoes. The magma also invades the crust to form granitic bodies called plutons, which exist in a variety of shapes and sizes. The largest plutons, called batholiths, form granitic mountains such as California's Sierra Nevada. Next in size are stocks and laccoliths, which form smaller, often isolated mountains composed of finer-grained granite. Sills and dikes form when magma invades weaknesses in the crust, such as sedimentary beds and fissures.

As mentioned, subduction zone volcanism builds volcanic chains on the continents and island arcs in the ocean. The rock type associated with these volcanoes is fine-grained, gray andesite, which contains abundant silica, suggesting a deep-seated source, possibly as much as 70 miles below the surface. Subduction zone volcanoes, such as those in Indonesia and in the western Pacific, are among the most explosive volcanoes in the world. Their violent nature is due to large amounts of volatiles, consisting of water and gases, in their magmas. As the magma rises toward the surface, the pressure drops, and volatiles escape explosively, shooting out of the volcano as though propelled by a gigantic cannon.

The Indonesian volcanoes Tambora and Krakatau, which are produced by the subduction of the Australindian plate down the Java Trench, are classic examples of subduction zone volcanism. The Alaskan volcanoes Katmai and Augustine are noted for their massive ash eruptions due to the subduction of the Pacific plate down the Aleutian Trench. And the 1991 highly explosive eruption of Pinatubo in the Philippines resulted from the subduction of the western Pacific plate down the Philippine Trench.

The Cascade Range in the Pacific Northwest is a chain of powerful volcanoes associated with the Cascadia subduction zone, which is being overridden by the North American continent. The 1980 eruption of Mount St. Helens, whose blast leveled 200 square miles of national forest, is a good example of the explosive nature of these volcanoes. The eruption marked the beginning of a new episode of eruptive activity in the Cascades and other parts of the world.

The volcanic rock associated with subduction zones, called andesite, contains a high percentage of silica. The rock derives its name from the Andes Mountains, whose volcanoes are highly explosive due to subduction of the Nazca plate down the Peru-Chile Trench. As the magma rises toward the surface, the molten rock feeds volcanic magma chambers and buoys up the Andes Mountain chain.

The second most common form of volcanism is rift volcanoes, which account for about 80 percent of all oceanic volcanism. Along spreading ridges, magma wells up from the upper mantle and spews out onto the ocean floor. The diverging lithospheric plates grow by the steady accretion of solidifying magma to their edges. Over 1 square mile of new oceanic crust composed of about 5 cubic miles of basalt is generated in this manner annually. At times, gigantic flows erupt on the ocean floor, producing massive quantities of new basalt.

Huge undersea fissure eruptions on the ocean floor at spreading ridges produce megaplumes of hot water. Short periods of intense volcanic activity can generate megaplumes measuring up to several tens of miles wide. The ridge splits open and spills out hot water, while lava erupts in an act of catastrophic seafloor spreading. In a matter of a few hours, or at most a few days, over 100 million cubic yards of superheated water gushes from a large crack in the oceanic crust up to several miles long. When the seafloor splits open in such a manner, it releases vast amounts of hot water held under great pressure beneath the surface.

Rift volcanoes, created by the divergence of lithospheric plates as the upper mantle is exposed to the surface, produce massive floods of basalt (Fig. 33). The East African Rift, which extends from eastern Mozambique to the Red Sea, is seething with volcanism. The rift is a complex system of tensional faults, indicating that the continent is in the

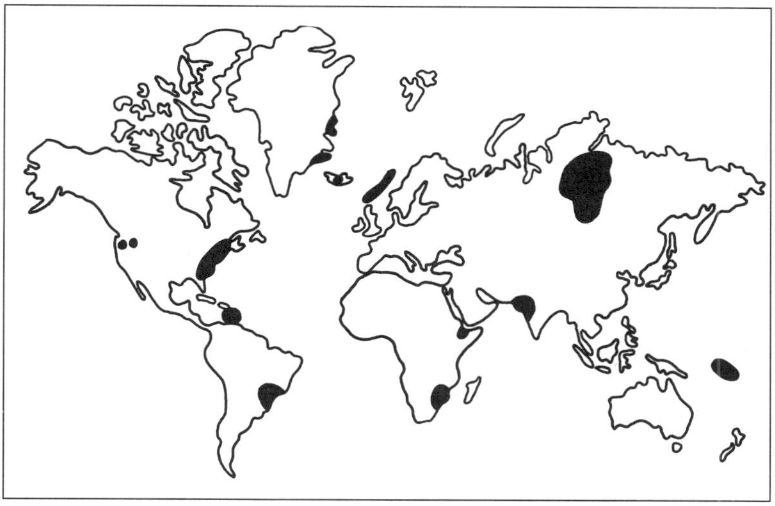

Figure 33   Areas affected by flood basalt volcanism.

initial stages of rupture. In the process of rifting, large earthquakes rumble across the landscape, as huge blocks of crust drop down along diverging faults. Much of the region has been uplifted thousands of feet by an

expanding mass of magma lying just beneath the thinning crust. This heat source is responsible for a number of volcanoes along the rift valley.

Iceland is a surface expression of the Mid-Atlantic Ridge. It is bisected by a volcanic rift that is responsible for the high degree of volcanism on the island. The rift produces a steep-sided, V-shaped valley flanked by several active volcanoes, making Iceland one of the most volcanically active places on Earth (Fig. 34). Often, volcanoes erupt under glaciers and melt the ice, causing massive floods to gush toward the sea. Geo-

Figure 34   A lava flow that has partially engulfed a building from the 1973 Heimaey eruption, Iceland. Courtesy of USGS

thermal energy produced by the volcanic activity provides over 90 percent of the heat for buildings, without which Iceland would be an unbearably cold place to live.

Another type of volcanism produces hot spot volcanoes, which exist in the interiors of plates far from plate margins where most volcanoes occur. They derive their magma from deep within the mantle, possibly as deep as the top of the core. The magma rises in giant mantle plumes that provide a steady flow of molten rock into magma chambers.

Perhaps the best example of hot spot volcanism is the volcanoes that built the Hawaiian Islands, which appear to have been created by a single mantle plume, beginning about 5 million years ago. The islands were assembled as though on a conveyor belt, with the Pacific plate traversing over the hot spot in a northwestward direction about 3 inches per year. Kilauea, whose name means much spreading, continues to erupt, and since 1983 its lava flows have covered almost 40 square miles, adding about 300 acres of new land to Hawaii.

Hot spots beneath continents produce domes or swells in the crust up to a hundred or more miles across, accounting for about 10 percent of the total land surface. As the dome grows, it develops deep fissures, through which magma finds its way to the surface. Africa is known for its many hot spots, which explains its unusual topography, consisting of numerous basins, swells, and uplifted highlands.

A hot spot under the United States is responsible for the volcanic activity in the Yellowstone Caldera, which covers an area about 45 miles long by 25 miles wide. Over the last 2 million years, three massive volcanic eruptions have devastated the region. The eruptions are counted among the greatest catastrophes of nature, and according to geologic history another major eruption is well overdue.

A single volcanic eruption can produce from a few cubic yards to as much as 5 cubic miles of volcanic material, composed of lava and pyroclastics (hot solid products) along with large quantities of water vapor and gases (Fig. 35). Rift volcanoes, which account for about 15 percent of the world's active volcanoes, generate about 2.5 billion cubic yards per

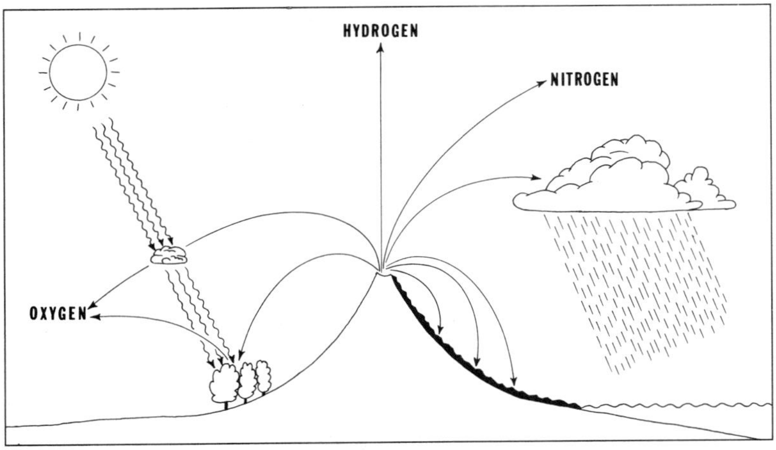

Figure 35 Volcanoes contribute large amounts of water vapor, carbon dioxide, and other important gases to the atmosphere, and also contribute to the growth of the continents.

Figure 36 The Mauna Loa Volcano, Hawaii. Courtesy of USGS

year, mainly submarine basalt flows. Some 20 eruptions of submarine rift volcanoes occur each year. Subduction zone volcanoes produce about 1 billion cubic yards of volcanic material per year, mostly pyroclastics. More than 80 percent of the subduction zone volcanoes, some 400 in all, exist in the Pacific. Hot-spot volcanoes produce about 500 million cubic yards per year, mostly basalt flows on the ocean floor, and pyroclastics and lava flows on the continents.

Volcanoes exist in a variety of shapes and sizes, depending on the type of eruption. Cinder cones are relatively short with steep slopes and form by explosive eruptions that deposit layer upon layer of pumice and ash. If a volcano erupts only lava from a central vent or fissure, it forms a broad shield volcano, like Mauna Loa (Fig. 36) on the big island of Hawaii, the largest of its kind in the world. The basaltic lava spreads out, covering as much as 1,000 square miles.

A volcano that erupts a composite of cinder and lava is called a stratovolcano, the tallest of volcanoes. Mount St. Helens is such a volcano. Highly explosive ash eruptions are followed by milder lava eruptions that reinforce the volcano's flanks, allowing it to grow to prodigious heights. Often these volcanoes end by decapitating themselves in a highly explosive eruption or by catastrophic collapse.

If such a cataclysm occurred at sea, it could generate a highly destructive tsunami. Volcanic eruptions that develop tsunamis are responsible for about a quarter of all deaths caused by tsunamis. The powerful waves can transmit the volcano's energy to areas outside the reach of the volcano itself. Large pyroclastic flows into the sea or landslides triggered by eruptions also can produce tsunamis. In 1792, during an earthquake after the eruption of Unzen, Japan, the side of the volcano collapsed into the bay, creating an enormous tsunami up to 180 feet high. It washed coastal cities out to sea, and up to 15,000 people vanished without a trace.

GAS EXPLOSIONS

In northwest Cameroon of central Africa, a region of volcanic peaks and valleys covered by lush tropical vegetation, a deep crater lake known as Lake Nios exploded on August 21, 1986. The gas eruption sent a deadly pall of toxic fumes—carbon dioxide, carbon monoxide, sulfur dioxide, hydrogen sulfide, and cyanide—spilling down the hillside. The gases spread out in a low hanging blanket over a distance of more than 3 miles downwind from the lake. The hot, humid gases clung to people's clothing, which they tried frantically to discard. The gases immediately asphyxiated almost all villagers, and 1,700 people along with thousands of cattle and other animals lost their lives. The gas eruption injured a total of 20,000 people over a 10-square-mile area.

The disaster might have been the result of an earth tremor that cracked open the floor of the lake, releasing volcanic gases under great pressures. The gas discharge created a huge bubble that burst through the surface of the lake, churning it to a murky reddish brown from the stirred-up bottom sediments. In addition, the temperature of the water had risen 10 degrees Celsius. A similar eruption two years earlier at Lake Mamoum in the same range killed 37 people, suggesting that the disaster at Lake Nios was no isolated occurrence.

On June 1, 1912, one of the 20th century's greatest eruptions produced a series of gigantic explosions that excavated a deep depression at the west base of Mount Katmai, Alaska, upon which a viscous lava rose pancake-shaped 800 feet in diameter and nearly 200 feet high. The lava covered an adjacent valley with a yellowish orange mass 12 miles long and 3 miles wide. Thousands of white fumaroles (volcanic steam vents) gushed from the lava flow, shooting hot water vapor as high as 1,000 feet into the air. Explorers discovering this wonder named the area the Valley of Ten Thousand Smokes.

Fumaroles are vents at the Earth's surface in volcanic regions that expel hot gases, often explosively. They exist on the surface of lava flows, in the calderas and craters of active volcanoes, and in areas where hot intrusive magma bodies occur. The gas temperatures within the fumarole can reach 1,000 degrees Celsius. The primary requirement for the production of fumaroles and geysers is that a large, slowly cooling magma body lies near the surface to provide a continuous supply of heat.

The hot water and steam originate from juvenile water released directly from magma and other volatiles or from groundwater percolating downward near a magma body that heats the water by convection currents. Volatiles released from the magma body also can heat the groundwater from below. Generally, the bulk of the gases consists of steam and carbon dioxide, with smaller amounts of nitrogen, carbon monoxide, argon, hydrogen, and other gases. In another type of fumarole, called a solfatara, from the Italian meaning sulfur deposit, sulfur gases predominate.

Another result of the explosive release of trapped gases is blowouts. Hole in the Ground in the Cascade Range in Oregon is the site of a gigantic volcanic gas explosion that created a huge crater. The crater is a nearly perfect circular pit that is several thousand feet across, with a rim raised several hundred feet above the surrounding terrain. For mysterious reasons, most of the crater lacks vegetation.

Pockets of gas also lie beneath the ocean floor, trapped under high pressure. As the pressure increases, the gases explode undersea and spread debris over wide areas, producing huge craters on the seabed. The gases rush to the surface in great masses of bubbles that burst when reaching the open air, resulting in a thick foamy froth on the ocean. In 1906, sailors in the Gulf of Mexico, southwest of the Mississippi River Delta, witnessed such a gas blowout that sent mounds of bubbles to the surface.

Further exploration of the site revealed a large crater on the ocean floor below 7,000 feet of water. The elliptical crater measured 1,300 feet long, 900 feet wide, and 200 feet deep and sat on a small submarine hill. Downslope from the crater was over 2 million cubic yards of ejected debris. Apparently, the gases seeped upward along cracks in the seafloor and collected under an impermeable barrier. Eventually, as the pressure grew, the gas blew off its cover, forming a huge blowout crater.

HAZARDOUS VOLCANOES

A number of volcanoes in the western United States are likely to erupt in the future. Some, like Mount St. Helens, might be awaking from long slumbers. Even 50,000 years of quietude might not be long enough to

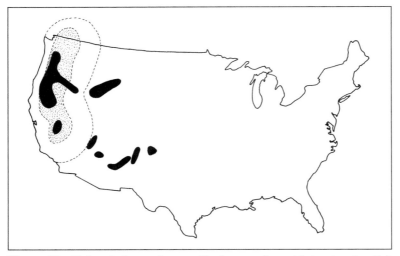

Figure 37   Volcanic hazard areas. Dark areas have highest potential, stippled area second highest, and dashed area third highest.

silence the rumblings of volcanoes, some of which have awakened even after a million years of sleep. Forecasting future eruptions requires a determination of a volcano's past behavior, revealed by studying its rocks. The volcano can then be grouped along with others by order of hazard.

More than 35 volcanoes in the United States, mostly in the Cascade Range, are likely to erupt sometime in the future. The most hazardous are volcanoes that have erupted on average every 200 years, that have erupted in the past 300 years, or both. These include, in descending hazardous order, Mount St. Helens, the Mono-Inyo Craters, Lassen Peak, Mounts Shasta, Rainier, Baker, and Hood. The next most hazardous are volcanoes that erupt less frequently than every 1,000 years and last erupted over 1,000 years ago. These include Three Sisters, Newberry Volcano, Medicine Lake Volcano, Crater Lake Volcano, Glacier Peak, Mounts Adams, Jefferson, and McLoughlin. The third most hazardous are volcanoes that last erupted more than 10,000 years ago but still overlie large magma chambers. These include Yellowstone Caldera, Long Valley Caldera, Clear Lake Volcanoes, Coso Volcanoes, San Francisco Peak, and Socorro, New Mexico.

A map of geologically recent eruptions shows 75 centers of volcanic activity arrayed in broad bands (Fig. 37). They extend from the Cascade Range in northern California, Oregon, and Washington eastward through Idaho to Yellowstone and along the

Figure 38   The May 22, 1915, eruption of Mount Lassen, Shasta County, California. Photo by B. F. Loomis, courtesy of USGS

border between California and Nevada. Another band extends from southeast Utah through Arizona and New Mexico. Because the pattern of activity 5 million years ago closely resembles the pattern since 10,000 years ago, all centers of activity have the potential for future eruptions, and new centers of activity might form within these bands any time.

Historical records indicate that before the 1980 eruption of Mount St. Helens, only two other eruptions had taken place in the Cascade Range in the last 90 years. A minor ash eruption occurred at Mount Hood, Oregon, in 1906, and several spectacular eruptions of Lassen Peak, California, occurred between 1914 and 1917 (Fig. 38). Between 1832 and 1880, Mounts Baker, Rainier, St. Helens, and Hood erupted ash or lava. Periods between eruptions were 10 to 30 years for each volcano, and perhaps as many as three volcanoes erupted in the same year. However, neither of these can compare to the eruptions of Mount St. Helens in its latest reawakening.

4

EARTH MOVEMENTS

All earth movements are naturally recurring events that have become increasingly hazardous due to human activities. Slopes are the most common and among the most unstable landforms. Under favorable conditions, the ground can give way even on the gentlest slopes. Material on most slopes is constantly on the move at rates varying from imperceptible creep of soil and rock to catastrophic landslides and rockfalls that travel at tremendous speeds, often destroying property and taking lives.

LANDSLIDES

Landslides are rapid downslope movements of soil and rock materials under the influence of gravity (Fig. 39), triggered mostly by earthquakes and violent weather systems. The main types of landslides are falls, topples, slides, spreads, and flows. Landslides consisting of overburden alone are called debris slides, which are the most dangerous form of slope movement with respect to human life.

Slides consisting of bedrock are called rockslides and slumps. Slumps develop where strong, resistant rock overlies weaker rocks. Material slides downward in a curved plane, tilting up the resistant unit, while the weaker rock flows out forming a heap. Slumps develop new cliffs just below pre-existing ones, setting the stage for renewed slumping. Slumping is a continuous process, and usually many previous generations of slumps exist far in front of the present cliffs. Therefore, all cliffs are inherently unstable and only temporary features over geologic time.

Figure 39   The June 23, 1925, Gros Ventre landslide, Lincoln County, Wyoming. Photo by W. C. Alden, courtesy of USGS

Landslides occur in earth materials that fail along planes of weakness under stress. They are initiated by an increase in stress and a reduction of strength in a rock formation, usually due to the addition of water to a slope. The slope geometry along with the composition, texture, and structure of the soil determines the formation strength. Changes in pore pressure and water content could weaken the friction between rock layers. The maximum natural inclination of a slope, called the angle of repose, is self-regulated by triggering slides that bring the slope back to its critical state when it becomes oversteepened.

Rock, soil, or snow particles pulled down a slope by gravity rub against each other and the ground as they fall. Each interaction causes particles to change direction and lose energy to friction. Usually, the shallower the slope angle the lower the friction within the flow. Particles on the bottom of the slide in contact with the bed slow down, while the upper particles glide over them in a tumbling, chaotic mass. In this manner, the flow resembles a dense gas of heavy, colliding particles rather than a fluid.

Individual landslides generally are not as spectacular as other violent forms of nature. However, they are more widespread, causing major economic losses and casualties in virtually every region of the world. In addition, landslides occur with other geologic hazards, including earthquakes, volcanic eruptions, and floods. The most severe landslides in the United States occur in mountainous regions, such as the Appalachian and Rocky mountains and ranges along the Pacific Coast.

The direct costs from damage to highways, buildings, and other facilities as well as indirect costs resulting from the loss of productivity exceed $1 billion annually. Single large slides in populated areas can cost tens of millions of dollars. Fortunately, landslides in the United States have not

Figure 40 A large rock slide near the summit of the Santa Cruz Mountains from the October 17, 1989, Loma Prieta, California, earthquake. Photo by G. Plafker, courtesy of USGS

resulted in a major loss of life like those in other parts of the world because most catastrophic slope failures in this country generally occur in sparsely populated areas.

One exception is California, whose residents are well familiar with landslides in their state (Fig. 40). Thousands of landslides causing considerable property damage have occurred in the Los Angeles basin alone. During the 1950s, repeated heavy rains and floods devastated the hillsides of Los Angeles, setting off landslides that destroyed or seriously damaged hundreds of homes. In response to this year-round destruction, officials began dealing with nonearthquake geologic hazards of the mountainous and hilly seaside regions by enacting landslide-control legislation that requires new building sites to be inspected by an accredited geologist.

Most landslides are triggered by earthquakes. The size of the area affected by these landslides depends on the magnitude of the tremor, the topography and geology of the ground, and the amplitude and duration of the ground motion. During the 1959 Hebgen Lake, Montana, earthquake, which killed 26 people, a single large slide, moving from north to south, gouged out a huge scar in a mountainside (Fig. 41). Debris traveled uphill on the south side of the valley and dammed the Madison River, creating a large lake.

Earthquake-induced landslides often result in wide area destruction. The 1971 San Fernando, California, earthquake produced nearly a thousand slides distributed over 100 square miles of remote and hilly mountainous terrain. The 1976 Guatemala City earthquake triggered some 10,000 landslides throughout an area of 6,000 square miles. On March 5, 1987, during the rainy season in Ecuador, an earthquake shook loose fierce mudslides that buried villages in the rugged hilly region, killing over 1,000 people.

Landslides also can be triggered by the removal of lateral support by erosion from streams, glaciers, waves, or currents. They are initiated by previous slope failures and human activities, such as excavation and other forms of construction. The ground then gives way under excess loading by the weight of rain, hail, or snow. In addition, the weight of buildings and other man-made structures tends to overload a slope causing it to fail.

The most common triggering mechanisms for landslides—after vibrations from earthquakes—are explosions that break the bond holding a slope together, overloading of a slope so that it can no longer support its new weight, undercutting at the base of a slope, and oversaturation with water from rain or melting snow. Water adds to the weight of a slope and decreases the internal cohesion of overburden. The effect of water as a lubricant is very limited, however. Its main effect is the loss of cohesion when the spaces between soil grains are filled with water.

Figure 41   The August 1959 Madison Canyon slide, Madison County, Montana. Photo by J. R. Stacy, courtesy of USGS

In volcanic mountainous regions, seismic activity and uplift associated with an eruption cause landslides in thick deposits of unconsolidated pyroclastic material on a volcano's flanks. The distribution of landslides associated with volcanoes is determined by the seismic intensity, topographic amplification of the ground motion, the rock type, slope steepness, and fractures and other types of weakness in the rock. Heavy sustained rainfall over a wide area also can trigger landslides and mudflows in volcanic terrain.

The longest landslide ever reported tumbled off Mexico's Nevado de Colima Volcano more than 18,000 years ago. The Colima slide sped 75 miles to the Pacific Coast and then some distance into the ocean. The western rim of Mount Etna's Valle del Bove caldera resulted from a collapse of the mountain's eastern slope centuries ago. Ground motion has destabilized the caldera, which could lead to rockslides, mudslides, or even a full-scale eruption.

Every 150 to 200 years, large parts of Alaska's Mount St. Augustine (Fig. 42) have collapsed and fallen into the sea, generating large tsunamis. Massive landslides have ripped out the flanks of the volcano ten or more times during the past 2,000 years. The most recent slide occurred during the October 6, 1883, eruption, when debris on the flanks of the volcano crashed into the Cook Inlet, sending a 30-foot tsunami to Port Graham 54 miles away that destroyed boats and flooded houses. Subsequent eruptions have filled the gap left by the last landslide, making the volcano increasingly unstable and setting the stage for another collapse. If a landslide does occur, it would barrel down the north side of the volcano and plunge into the sea, sending a tsunami in the direction of cities and oil platforms residing in the inlet.

Another type of downslope movement is volcanic pyroclastic flows, which are masses of hot dry rock debris that move rapidly down a volcano's flanks like a fluid. Their mobility is due to hot air and other gases mixed with the debris called a nu'ée ardente. They often form when large masses of hot rock fragments suddenly erupt onto a volcano's flanks. Pyroclastic flows travel at speeds up to 100 miles per hour and tend to follow valley floors, burying them with thick deposits. Because of their great mobility, pyroclastic flows can affect areas 15 miles or more away from a volcano. The swiftly moving hot rock debris buries and incinerates everything in its

Figure 42   Mount St. Augustine, Kamishak district, Cook Inlet, Alaska. Photo by C. W Purington, courtesy of USGS

path. Clouds of dust and gas can blanket adjacent areas downwind, burning and asphyxiating people and animals.

Landslides are often called avalanches, but this term is generally reserved for snow slides. Avalanches usually begin with a mass of fresh, powdery snow resting on a steep bank of older snowpack and are triggered by loud noises, earthquakes, or skiers. A spectacular example of an avalanche took place in the Andes Mountains of Peru on May 31, 1970. A large earthquake triggered a sliding

Figure 43 A large boulder transported by the May 31, 1970, avalanche in the Andes Mountains, Peru. Courtesy of USGS

mass of glacial ice and rock, 3,000 feet wide and about a mile long, that rushed rapidly downslope with a deafening roar. Frictional heat partially melted the ice, making the slopes even more slippery. The avalanche traveled nearly 10 miles to the town of Yungay in 4 minutes or less, burying it under thousands of tons of rubble.

The trajectories of thousands of boulders, weighing up to several tons and hurled more than 2,000 feet across the valley (Fig. 43), indicated the slide reached a velocity of nearly 250 miles per hour. The slide shot across the valley and 175 feet up the opposite bank, where it partly destroyed another village. Flash flooding from broken mountain lake basins and from the avalanche-swollen waters of the Rio Santa River created a wave as high as 45 feet, which caused flooding that only exacerbated the disaster. When it was all over, 18,000 people had lost their lives.

ROCKSLIDES

Rockslides are generally large and destructive, often involving millions of tons of rock, created by a mass of bedrock that breaks into many fragments during its fall. The material behaves like a fluid, spreading out on the floor below. The slide might have enough energy to flow some distance uphill on the opposite side of a valley. Rockslides are prone to develop when planes of weakness, such as bedding planes or jointing, are parallel to a slope. This is especially true if the slope has been undercut by a river, glacier, or construction work.

One of the most destructive rockslides took place on October 9, 1963, at Monte Toc in the Italian Alps, named by the local residents as "the mountain that walks." Despite efforts to stabilize the slopes, the mountain not only walked, it galloped. A torrent of water, mud, and rock plunged into the narrow gorge, shot across the Piave River, and ran up the mountain slope on the opposite side. The slide completely demolished the town of Longarone, killing 2,000 of its residents. Some 600 million tons of debris slid into a reservoir, forcing the water 800 feet above its previous level. The water rose in one great wave 300 feet above the dam and dropped into the gorge below. The water quickly picked up speed, snatching tons of mud and rock as it raced on its destructive journey downstream.

On September 11, 1881, the town of Elm, Switzerland, was wiped off the map when a nearby mountainside suddenly collapsed, transforming a solid cliff into a river of rock. The collapsing cliffside plummeted 2,000 feet and sped through the valley below for a distance of nearly 1.5 miles. As the gigantic mass of rock debris rapidly roared down the valley, it entombed 116 people beneath a thick blanket of broken slate before grinding to a halt. Soon after the slide, the Swiss geologist Albert Heim visited the site, where he discovered that such large slides, called long-runout landslides, travel for long horizontal distances because they encounter little base-level friction to slow them down.

If material drops off a nearly vertical mountain face at the velocity of free-fall, it results in a rockfall or soilfall, depending on its composition. Rockfalls range in size from individual blocks plunging down a mountain slope to the failure of huge masses of rock weighing hundreds of thousands of tons falling nearly vertically down a mountain slope. Individual blocks commonly come to rest at the base of a cliff in a loose pile of angular blocks, called talus or scree (Fig. 44).

Immensely destructive waves are set in motion if large blocks of rock drop

Figure 44   Large talus cones in Stinking Water Canyon, Park County, Wyoming. Photo by Jaggar, courtesy of USGS

Figure 45   The trimline shown in light areas, where trees were destroyed by an enormous rockslide into Lituya Bay, Alaska, in 1958. Courtesy of USGS

into a standing body of water, such as a lake or fjord. A 1958 earthquake in Alaska triggered an enormous rockslide that fell into Lituya Bay, generating a gigantic wave that surged 1,700 feet up the mountainside. Trees toppled over like matchsticks when massive amounts of seawater inundated the shores (Fig. 45). Coastal landslides of large magnitude also can generate destructive tsunamis. This hazard is particularly feared in Norway, where small deltas might provide the only available flat land. Waves generated by rockfalls can range from 20 to 300 feet high and cause considerable damage as they burst through local villages.

The most impressive rockfall ever recorded occurred at Gohna, India, in 1893. A huge mass of rock, loosened by driving monsoon rains, dropped 4,000 feet into a narrow Himalayan Mountain gorge, forming a gigantic dam 3,000 feet across and 900 feet high. The huge pile of broken rock, measuring about 5 billion cubic yards, impounded a large lake over 700 feet deep. Two years after the fall, the dam burst, causing a world record flood. Ten billion cubic feet of water discharged within hours, producing floodwaters that crested 240 feet high.

The largest rockfall in the Rocky Mountain region in several thousand years occurred in Alberta, Canada, on April 29, 1903. It might have been triggered by coal mining below the base of Turtle Mountain. A mass of strongly jointed limestone blocks on the crest of the mountain broke loose and plunged down the deep escarpment. About 90 million tons of material tumbled down the mountainside and swept through the town of Frank in one tremendous wave, killing 70 people.

SOIL SLIDES

Earthquakes can cause soil slides involving weakly cemented, fine-grained materials that form steep, stable slopes. The 1920 Kansu, China, earthquake produced soil-flow failures that killed an estimated 180,000 people. As the tremor rumbled through the region, immense slides rushed out of the hills, burying entire villages and damming streams that flooded valleys.

Soil on a steep hillside can suddenly change into a wave of sediment, sweeping downward at speeds of over 30 miles per hour. Precipitation frees dirt and rocks by increasing the water pressure in pores within the soil. As the water table rises and pore pressure increases, friction holding the top layer of soil to the hillside diminishes until the pull of gravity overcomes it. Immediately before the soil begins to slide, the pore pressure drops and the soil starts to expand.

Creep (Fig. 46) is a slow downslope movement of soil, often recognized by downhill-tilted poles, fence posts, and trees. It is a more rapid movement of near-surface soil material than the sediment below and is particularly rapid where frost action is prominent. After a freeze-thaw sequence, material moves downslope due to the expansion and contraction of the ground. Under these conditions, trees are unable to root themselves, and only grasses and shrubs can grow on the slope. If creep is especially slow, tree trunks are bent, and after the trees become tilted new growth attempts to straighten them. If the creep is continuous, trees lean downhill in their lower parts and become progressively straighter higher up.

An earthflow (Fig. 47) is a more visible form of movement caused by raised water content in overburden, which increases the weight and reduces the stability of the slope by lowering resistance to shear. Earthflows are characterized by grass-covered, soil-blanketed hills, and although they generally are minor features some can be considerably large, covering several acres. Earthflows usually have a spoon-shaped sliding surface, upon which a tongue of overburden breaks away and flows for a short distance, forming a curved scarp at the breakaway point.

Figure 46   Railroad track bent by creep in the Nome River Valley, Alaska. Photo by R. S. Sigafoos, courtesy of USGS

Expansive soils are sediments that swell or shrink due to changes in moisture content. Buildings and other structures built on expansive soils cost the United States several billion dollars annually in damages. Expansive soils are abundant in geologic formations in the Rocky Mountain region, the Basin and Range Province, the Great Plains, the Gulf Coastal Plain, the lower Mississippi River Valley, and the Pacific Coast. The parent materials for expansive soils originate from volcanic and sedimentary rocks that decompose into clay minerals that tend to form highly unstable slopes.

Figure 47   An earthflow on slopes west of Prosperity, Washington County, Pennsylvania. Photo by J. S. Pomeroy, courtesy of USGS

MUDFLOWS

Mudflows are among the most impressive features of the world's deserts (Fig. 48). Heavy runoff can form rapidly moving sheets of water that pick up huge quantities of loose material. The floodwaters flow into a stream, where all the muddy material suddenly concentrates. The dry streambed rapidly transforms into a flash flood that moves swiftly downhill, often with a steep, wall-like front. The behavior of mudflows is similar to that of a viscous fluid, often including a tumbling mass of rocks and large boulders.

Figure 48   Alluvial fans from outwash flows on the floor of Death Valley, Inyo County, California. Photo by H. E. Malde, courtesy of USGS

Mudflows can cause considerable damage as they flow out of mountain ranges. At the base of the range, velocity drops, and the loss of water by percolation into the ground thickens the mudflow until it comes to a complete halt. Mudflows can carry large blocks and boulders the size of automobiles onto the floor of desert basins far beyond the base of the bordering mountain range. Often, huge monoliths rafted out beyond the mountains by swift-flowing mudflows are stranded on the desert floor. Heavy rains falling on loose pyroclastic material on the flanks of volcanoes also produce mudflows.

Lahars are mudflows produced by volcanic eruptions. They derive their name from an Indonesian word meaning "mudflow" due to their common occurrence in this region. A tragic example is the 1919 eruption of Kelut Volcano on Java, which blew out the crater lake at its summit and created a large mudflow that killed 5,000 people. Lahars are masses of water-saturated rock debris that descend the steep slopes of volcanoes similar to the flowage of wet concrete. The debris comes from loose, unstable rock deposited on the volcano by explosive eruptions. The water is derived from rain, melting snow, a crater lake, or a reservoir next to the volcano. Lahars also can be initiated by a pyroclastic or lava flow moving across a glacier rapidly melting it. This happened during the May 18, 1980, eruption of Mount St. Helens, which created many destructive mudflows (Fig. 49).

A lahar's speed depends mostly on its fluidity and the slope of the terrain. Lahars can travel swiftly down valley floors for a distance of up to 50 miles or more at speeds exceeding 20 miles per hour. Lava flows extending onto glacial ice or snowfields produce floods as well as lahars. Flood-hazard zones extend long distances down some valleys. For volcanoes in the western Cascade Range, flood-hazard zones reach as far as the Pacific Ocean. Losses from lahars decrease rapidly with increasing height above the valley floor and gradually with increasing distance from the

Figure 49   Area affected by massive landslides and mudflows from the May 18, 1980, Mount St. Helens eruption, showing Spirit Lake in the foreground. Photo by Jim Hughes, courtesy of USDA–Forest Service

volcano. The vast carrying power of lahars can easily sweep away people and their structures.

The most tragic example of a mudflow initiated by a volcanic eruption in recent history was the November 13, 1985, eruption of Navado del Ruiz in Colombia. The eruption melted the volcano's ice cap and sent floods and mudflows cascading down the mountainside into the nearby Lagunilla and Chinchina river valleys. The mudflow had a consistency of mixed concrete and carried off everything in its path. It buried almost all of Armero, a city 30 miles from the volcano, and badly damaged 13 smaller towns, killing 25,000 people and leaving 60,000 homeless.

SUBMARINE SLIDES

Some of the largest and most damaging slides occur on the ocean floor, and 40 giant submarine slides have been located around U.S. territory. Submarine slides moving down steep continental slopes have buried undersea telephone cables under a thick layer of rubble. A modern slide that broke submarine cables near Grand Banks, south of Newfoundland, moved downslope at a speed of about 50 miles per hour. During the 1964 Good Friday Alaskan earthquake, submarine slides carried away large sections of the port facilities at Whittier, Valdez, and Seward.

Submarine flow failures can generate large tsunamis that overrun parts of the coast. For example, on July 3, 1992, what appeared to be a large undersea slide sent a 25-mile-long, 18-foot-high wave crashing down on Daytona Beach, Florida, overturning automobiles and injuring 75 people. In 1929, an earthquake on the coast of Newfoundland set off a large undersea landslide and triggered a tsunami that killed 27 people.

Coastal landslides occur when a sea cliff is undercut by wave action and falls into the ocean. Direct wave attack at the base of a cliff quarries out weak beds and eventually undercuts the cliff until the overlying unsupported material collapses into the sea. Excessive rainfall along the coast also can lubricate sediments, causing huge blocks to slide into the ocean.

Undersea slides carve out deep submarine canyons in continental slopes. The slides consist of sediment-laden water that is considerably denser than the surrounding seawater, allowing sediments to move swiftly along the ocean floor. These muddy waters, called turbidity currents, can move down the gentlest slopes and transport immensely large blocks. Turbidity currents are also initiated by river discharge, coastal storms, or other currents. They deposit huge amounts of sediment that build up the continental slopes and the smooth ocean bottom below.

The continental slopes incline as much as 60 to 70 degrees and plunge downward for thousands of feet. Sediments that reach the edge of the continental shelf slide off the continental slope by the pull of gravity. Huge masses of sediment cascade down the continental slope by gravity slides that can gouge out steep submarine canyons and deposit great heaps of sediment. They are often as catastrophic as terrestrial slides and can move massive quantities of sediment downslope in a matter of hours.

The submerged deposits near Hawaii rank among the largest landslides on Earth. By far, the largest example of an ancient undersea rockslide occurred along the flank of a Hawaiian volcano. The slide measured roughly 1,000 cubic miles in size and spread some 125 miles from its point of origin. The collapse of the island of Oahu sent debris 150 miles across the deep-ocean floor, churning the sea into gargantuan waves. When part of Mauna Loa Volcano collapsed and fell into the sea around 100,000 years ago, it created a tsunami 1,200 feet high that was not only catastrophic for Hawaii but might even have caused damage along the coast of California.

At the bottom of the rift valley of the Mid-Atlantic Ridge in the middle Atlantic lies the remains of a massive undersea avalanche in 10,000 feet of water that surpasses in size any landslide in recorded history. One side of

Figure 50A   Severe gully erosion on a pasture in Shelby County, Tennessee, as a result of cultivating unsuited soil that should have been left fallow. Photo by Tim McCabe, courtesy of USDA Soil Conservation Service

the volcanic mountain range apparently gave way and slid downhill at a tremendous speed, running up and over a smaller ridge farther downslope in a matter of minutes. The slide carried about 4.5 cubic miles of rock debris, which is six times greater than the 1980 Mount St. Helens landslide, the largest in recorded history. The slide probably occurred around half a million years ago and possibly created a tsunami 2,000 feet high.

SOIL EROSION

Perhaps the greatest limitation to further human population growth is soil erosion (Figs. 50A & 50B). Soil is one of our most endangered resources. Not only does soil erosion rob fields of precious nutrients, it also increases farming costs and lowers food production. Prior to the advent of agriculture, some 6,000 years ago, natural soil erosion rates were probably no more than 10 billion tons per year, slow enough for new soil to be generated in its place. Today, however, the estimated soil erosion rates are about 25 billion tons per year. In other words, we are losing soil more than twice as fast as nature is putting it back. As much as one-third of the global cropland is losing soil at a rate that is undermining any long-term agricultural productivity. World food production per capita could eventually fall if the loss of topsoil continues.

Erosion rates vary depending on the amount of precipitation, the topography of the land, the type of rock and soil, and the amount of vegetative cover. Efforts to increase worldwide crop production through deforestation, irrigation, use of artificial fertilizers, genetic engineering, and other scientific methods could ultimately fail if the topsoil disappears. People would require a nearly 2 percent annual increase in food production to meet the global demand as the population continues to explode. Most of this increase would have to be made by new technology, especially since a significant portion of the world's cropland is already lost.

Many rivers, particularly those in Africa, which has the worst erosion problem in the world,

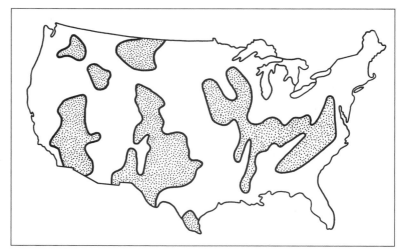

Figure 50B   Areas affected by soil erosion.

are becoming heavily sedimented due to topsoil erosion. The Mississippi River, which drains America's heartland, dumps more than a quarter billion tons of sediment into the Gulf of Mexico each year. In the United States, eroding cropland is costing nearly a billion dollars annually because of polluted and sedimented rivers and lakes. The sediments also severely limit the life expectancy of dams built for water projects such as irrigation. Therefore, the best way to control silt buildup is by adopting effective soil-conservation measures in the watershed so that less topsoil is lost to erosion.

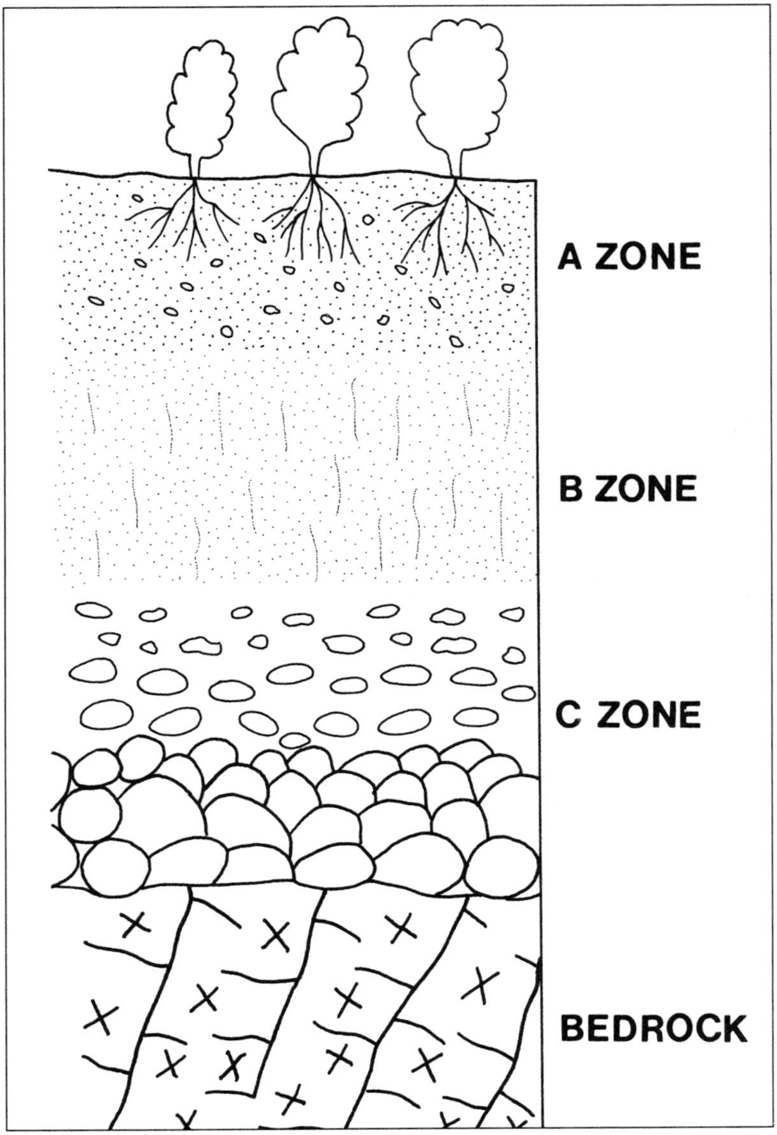

The soil profile (Fig. 51) begins with the A zone, which contains most of the soil nutrients. It is a thin bed from a few inches to a few feet thick, with an average thickness of 7 inches worldwide. Below lies the B zone, which is coarser and of poorer soil quality. As the A zone thins out and erosion brings the B zone to the surface, the potential for runoff and erosion is significantly increased because the B zone is generally unfavorable for sustaining vegetation, whose roots help hold the soil in place.

To keep up with an ever-growing human population, which adds another 100 million mouths to feed each year, farmers have abandoned sound soil conservation practices in favor of more intensified farming methods. This includes less rotation of crops,

Figure 51 The soil profile. A zone—organic rich, B zone—organic poor, C zone—parent rock material.

greater reliance on row crops, more plantings between fallow periods, and extensive use of chemical fertilizers rather than natural organic fertilizers that help bind the soil.

Over the last 150 years, intensive agriculture has reduced the average soil depth in the United States by about half. During the 1980s, U.S. cropland shrank by 7 percent, and presently between 2 and 4 billion tons of topsoil are lost annually. Due to the 1988 drought, for the first time since the Second World War, Americans consumed more food than they grew. Despite expected rises in temperatures, increased evaporation rates, and changes in rainfall patterns, the United States should still be able to produce adequate food to feed itself. Unfortunately, the nation's ability to produce excess food for export to help feed the world's hungry could be very limited.

Throughout the world, most of the arable land is already under cultivation, and efforts to cultivate substandard soils are leading to poor productivity and ultimately abandonment, which in turn leads to severe soil erosion. Marginal lands, which are often hilly, dry, or contain only thin, fragile topsoils and therefore erode easily, are also forced into production. As world populations continue to grow geometrically on a planet whose resources are dwindling rapidly, the vast majority of people could face a cultivation catastrophe.

5

CATASTROPHIC COLLAPSE

In this century, human activity has produced extensive and alarming geologic effects. Among the most far-reaching is the sinking of the land due to subsidence. This is a growing problem, worsening as people draw ever more deeply on stocks of water and petroleum. Such ground failures result from the dissolution of soluble materials underground or the withdrawal of fluids from subsurface sediments. Ground failures also occur when subterranean sediments liquefy during earthquakes or violent volcanic eruptions, causing considerable damage to man-made structures.

THE SINKING EARTH

The 1811 and 1812 New Madrid, Missouri, earthquakes, three of the greatest tremors in American history, caused major changes in ground levels over large areas. The town itself, which was totally demolished, subsided more than 12 feet, and new lakes littered with drowned cypress trees filled the basins of downdropped crust. Collapsing riverbanks changed the course of the Mississippi River toward the west, and whole islands disappeared, while new ones emerged elsewhere. The saturated bottomland soil spurted huge geysers of sand and black water a hundred or more feet into the air, forming craters up to 30 feet wide.

During the April 18, 1906, San Francisco, California, earthquake, the ground subsided several inches under buildings, causing them to collapse (Fig. 52). The damage was severest in the low-lying business district, where the earthquake destroyed or structurally weakened almost all buildings in the downtown area. The subsidence ruptured water mains, and fire fighters stood by helplessly as most of the city burned to the ground. Buildings that managed to survive the earthquake were utterly destroyed by fire.

The March 27, 1964, Good Friday Alaskan earthquake was the largest recorded on the North American continent. The estimated area of destruction was 50,000 square miles, and massive landslides and subsidence dev-astated large sections of

Figure 52 The April 18, 1906, San Francisco, California, earthquake caused these houses to shift to the left, while the tall house dropped from its south foundation wall and leaned against its neighbor. Photo by G. K. Gilbert, courtesy of USGS

Anchorage (Fig. 53). Thirty blocks were destroyed when the city's slippery clay substratum slid toward the sea. Landslides caused $50 million in damage, and houses were destroyed when 200 acres were carried seaward. The ground beneath Valdez, Seward, and Whittier gave way, causing large sections of port facilities to slide toward the sea.

Niigata, Japan, in the northwestern part of the main island of Honshu, overlies a large deposit of natural gas dissolved in saltwater. In order to

Figure 53   The Turnagain Heights landslide from the March 27, 1964, Alaskan earthquake.   Courtesy of USGS

extract the gas, large quantities of water had to be pumped out of the ground, producing an alarming amount of subsidence. Parts of the city sank below sea level, requiring the construction of dikes to keep the sea out. On June 16, 1964, a major earthquake struck the area. It breached the dikes and caused the city to subside a foot or more, resulting in serious flooding in the area of subsidence.

In Mexico City, over-pumping of groundwater has caused the Western Hemisphere's largest metropolis to subside over 20 feet since 1940. The discharge of water wells far exceeds the natural recharge of the aquifer beneath the city, reducing the fluid pressure as the water table lowers. Some areas have subsided more than a foot a year, often resulting in numerous earth tremors. The constant shaking of the ground might explain why residents ignored the foreshocks that preceded the destructive September 19, 1985, earthquake that destroyed large portions of the city and killed as many as 10,000 people.

GROUND FAILURES

Ground failures in water-saturated subterranean sediments during violent earthquakes and volcanic eruptions result from a process known as liquefaction. When seismic shear waves from an earthquake pass through a loose, saturated granular soil layer, they distort the structure and cause void spaces to collapse in loosely packed sediments. Each collapse places stress on the pore water surrounding the grains, which disrupts the soil and increases pressure of the pore water, causing it to drain. If drainage is restricted, the pore water pressure builds until it equals the pressure exerted by the weight of the overlying ground. Grain contact stress is temporarily lost and the granular soil layer flows like a fluid.

Liquefaction applies to certain geologic and hydrologic environments, particularly areas where sediments were deposited since the last ice age, or during the last 10,000 years, and where groundwater lies near the

Figure 54 Sand boils in an irrigated field near Hollister from the October 17, 1989, Loma Prieta, California, earthquake. Photo by G. Plafker, courtesy of USGS

surface, usually within 30 feet. Generally, the younger and less consolidated the sediments and the shallower the water table, the more susceptible the soil is to liquefaction, which usually occurs during earthquakes of magnitude 6 or greater. The potential for disaster is enormous because many of the world's major cities are partly built on these young sediments.

Sand boils often develop during the liquefaction process. They are fountains of water and sediment that spout upwards of 100 feet or more from the pressurized liquefied zone. Earthquakes can turn a solid, water-saturated bed of sand underlying less permeable surface layers into a pool of pressurized liquid that seeks its way to the surface. Water laden with sediment vents to the surface by artesianlike water pressures developed during liquefaction. Sand boils also can cause local flooding and the accumulation of large deposits of sediment (Fig. 54).

Ground failures associated with liquefaction include lateral spreads, flow failures, and loss of bearing strength. Lateral spreads are lateral movements of large blocks of soil in a subsurface layer during earthquakes (Fig. 55). Lateral spreads usually break up internally, forming fissures and

Figure 55   Lateral spread on Heber Road, showing associated displacements, fissures, and sand boils, from the October 15, 1979, Imperial Valley, California, earthquake. Photo by C. E. Johnson, courtesy of USGS

scarps. They generally develop on gentle slopes of less than 6 percent grade. Horizontal movements on lateral spreads can be as great as 10 to 15 feet, but where slopes are particularly favorable and the duration of the tremor is long, lateral movement can extend up to 100 feet or more.

During Alaska's 1964 Good Friday earthquake, lateral spreading of floodplain deposits near river channels damaged or destroyed over 200 bridges. The lateral spreads compressed bridges over the channels, buckled decks, thrust sedimentary beds over abutments, and shifted and tilted abutments and piers. Lateral spreads are also destructive to underground structures such as pipelines. During the 1906 San Francisco earthquake, several major water main breaks hampered fire fighting efforts. The inconspicuous ground failure displacements of up to 7 feet were largely responsible for the destruction of San Francisco (Fig. 56). To prevent this occurrence in the future, San Francisco was rebuilt with duplicate water mains, so that if one line ruptured during an earthquake water could be shut off and rerouted to another.

Flow failures are the most catastrophic type of ground failure associated with liquefaction. They consist of soil or blocks of intact material riding on a layer of liquefied sediments, usually over a distance of several tens of feet. However, under certain geographic conditions, they can travel at great speeds over distances of up to several miles. Flow failures usually develop in loose saturated sands and silts on slopes of greater than 6 percent grade and originate on land and on the seafloor near coastal areas. The 1920 Kansu, China, earthquake induced several massive flow failures that killed as many as 180,000 people. The 1964 Good Friday earthquake in Alaska produced submarine flow failures that destroyed seaport facilities. The flow failures also generated large tsunamis that overran coastal areas and caused additional damage and casualties.

When the ground supporting buildings and other structures liquefies and loses bearing strength, large deformations occur within the soil, causing settlement or collapse. Soils that liquefy under buildings induce bearing failures that cause them to subside or tip over. Deformation usually occurs whenever a layer of saturated, cohesionless sand or silt extends from near the surface to a depth of about the width of the building. The most spectacular example of this type of ground failure occurred during the June 16, 1964, Niigata, Japan, earthquake, when several multi-story apartment buildings tilted as much as 60 degrees. Most of the buildings were jacked back into an upright position and underpinned with piles.

Earthquakes also can cause certain clays, called quick clays, to lose strength and fail. Quick clay is composed of flakes of clay minerals arranged in very fine layers, with a water content of 50 percent or more. Under ordinary conditions, quick clay is a solid that can support a weight of over a ton per square foot of surface area. However, the slightest jarring from an earthquake can immediately turn it into a liquid. The large landslides in Anchorage, Alaska, during the 1964 Good Friday earthquake resulted from the failure of layers of quick clay along with other beds of saturated sand and silt. The severity of the earthquake was responsible for the loss of strength in the clay layers and liquefaction in the sand and silt

Figure 56   San Francisco in flames after the April 18, 1906, earthquake due largely to broken water mains and the unavailability of water for fire fighting efforts. Photo by T. L. Youd, courtesy of USGS

Figure 57   The east part of the Turnagain slide in Anchorage from the March 27, 1964, Alaskan earthquake. Courtesy of USGS

layers, which were the major causes of ground failures that destroyed much of the city (Fig. 57).

Severe subsidence also occurs in permafrost regions of the higher latitudes. Solifluction is the slow downslope movement of water-logged sediments that causes ground failures in colder climates. When frozen ground melts from the top down, during spring in the temperate regions or during summer in permafrost regions, it causes the soil to glide downslope over a frozen base. Solifluction can create many construction problems, especially in areas of permafrost. Foundations must extend down to the permanently frozen layers, or entire buildings might be damaged by the loss of support or by lateral movement downslope.

SUBSIDENCE

Subsidence is the downward settling or collapse of the land surface either locally or over broad regional areas without appreciable horizontal movement. It is mostly caused by the withdrawal of fluids or by shock waves from earthquakes and is a common problem in nearly every state in the Union. Many parts of the world have been steadily sinking due to the withdrawal of large quantities of groundwater or petroleum. Generally, the amount of subsidence is on the order of one foot for every 20 to 30 feet of lowered water table. Underground fluids fill intergranular spaces and support sediment grains. The removal of large volumes of fluid results in a loss of grain support, a reduction of intergranular void spaces, and the compaction of clays, which causes the land surface to subside wherever widespread subsurface compaction occurs (Fig. 58).

The extraction of natural gas or oil from the ground also can cause earthquakes. Low-magnitude, shallow earthquakes have occurred near gas and oil fields in the United States since the 1920s. The tremors rarely exceed magnitude 4, but they can buckle pipelines and shear off wellheads,

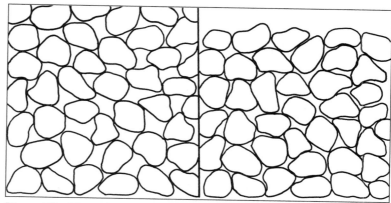

causing oil spills. Pumping oil out of the ground contracts the rock in the reservoir and sets up large pressure changes over short distances. Vertical contraction pulls down on the overlying rock and causes the ground above the reservoir to sink, and horizontal stresses pull the surrounding rock inward, causing the reservoir to contract like a drying sponge. If the pull is strong enough to shear the rock, it can trigger a

Figure 58   The subsidence of sediments (right) by the withdrawal of fluids.

mild earthquake. Furthermore, pumping petroleum from the ground can collapse the oil-filled fractures, which seals the remaining reservoir and causes oil wells to go dry.

The opposite condition occurs by pumping waste fluids into wells, which also triggers earth tremors. On January 31, 1986, a moderate earthquake struck 25 miles northeast of Cleveland, Ohio, the largest ever reported in the area. Since 1974, some 250 million gallons of hazardous liquids were pumped into two 5,900-foot-deep wells, and the resulting pressures enlarged fractures in the rock that might have activated a nearby fault. In the 1960s, a series of earthquakes hit Denver, Colorado, due to wastewater injection at the Rocky Mountain Arsenal. The liquids tend to unlock faults, causing rocks with pent up pressures to slip. Oil forced out of reservoirs using water injection wells also can cause minor earthquakes. Some 300,000 such wells exist in the United States today.

The most dramatic examples of subsidence in the United States occur along the Gulf Coast of Texas, in Arizona, and in California. The Houston-Galveston area has experienced local subsidence as much as 7.5 feet, and an average of a foot or more over an area of 2,500 square miles, mostly due to the withdrawal of large amounts of groundwater. In Galveston Bay, the ground subsided 3 feet or more over an area of several square miles due to the rapid pumping of oil from the underlying strata. Subsidence in some coastal towns has increased their susceptibility to flooding during severe coastal storms by dropping them closer to sea level.

At Long Beach, California, the ground subsided, forming a huge bowl up to 26 feet deep over an area of 22 square miles due to the withdrawal of large quantities of oil during the 1940s and 1950s. In some parts of the oil field, the affected land subsided at a rate of 2 feet per year. The downtown area subsided upwards of 6 feet, causing about $100 million in damage to the city's infrastructure. Most of the subsidence was halted by injecting seawater under high pressure into the underground reservoir, which fortuitously increased the production of the oil field by forcing the petroleum toward the surface.

Large areas of California's San Joaquin Valley have subsided because of intense pumping of groundwater for agricultural purposes. The arid region is so dependent on groundwater it accounts for about one-fifth of all well water pumped in the United States. The ground is sinking at rates of up to a foot a year. In some areas, the land has fallen more than 20 feet below former levels. Other parts of the San Joaquin Valley have subsided due to shrinking of the soil as it dries out. In the northern part of the valley, subsidence has dropped the land surface more than 10 feet below sea level, requiring the construction of protective dikes to prevent flooding.

Venice, Italy, is drowning because of a combination of rising sea levels and subsidence. High tides have increased in magnitude and frequency since 1916. Over the last 50 years, the cumulative subsidence of Venice has been about 5 inches. Meanwhile, the Adriatic Sea has risen about 3.5 inches over this century, resulting in a change of more than 8 inches between Venice and the sea. The city floods during high tides, heavy spring runoffs, and storm surges. The subsidence results from the overuse of groundwater, which causes compaction of the aquifer, or water-carrying strata, beneath the city. To stop the flooding, locks are needed to keep the sea from spilling into the lagoon upon which the city was built. Unfortunately, such a move could silt up the lagoon, causing Venice's famous canals to go dry. The city would no longer sit at the edge of the sea like it has for centuries.

In the northeastern section of Tokyo, Japan, the land has sunken around building foundations due, again, to overdrawing of groundwater. The subsidence progressed at a rate of half a foot a year over an area of about 40 square miles, 15 square miles of which sank below sea level, prompting the construction of dikes to keep out the sea from certain sections of the city. A threat of catastrophe hangs over Tokyo because it could well be inundated by floodwaters during a typhoon or an earthquake.

Subsidence from the withdrawal of groundwater can produce fissures, resulting in the formation of open cracks in the ground. Subsidence also can cause the renewal of surface movement in areas cut by faults. Surface fissuring and faulting resulting from the withdrawal of groundwater is a potential problem in the vicinity of Las Vegas, Nevada, as well as the arid regions of California, Arizona, New Mexico, and Texas. The withdrawal of large volumes of water and oil can cause the ground to subside to considerable depths, often with catastrophic consequences.

CATASTROPHIC COLLAPSE

The ground also can seriously collapse during earthquakes. Earthquake-induced subsidence in the United States has occurred mainly in California, Alaska, and Hawaii. The subsidence results from vertical displacements along faults that can affect broad areas. The town of New Madrid, Missouri, was totally demolished when the ground beneath it collapsed from a height of 25 feet to 12 feet above the Mississippi River during the massive earthquakes of 1811–1812. Two lakes formed in the basins of the downdropped crust, the largest of which is the 50-foot-deep Reelfoot Lake (Fig. 59). During the 1906 San Francisco earthquake, the ground subsided several inches under buildings, causing them to collapse. During the 1964 Good Friday earthquake, subsidence destroyed large sections of Anchorage, Alaska, when the clay substratum slid out toward the sea.

The earthquake also forced over 70,000 square miles of land to tilt downward more than 3 feet, causing extensive flooding in coastal areas of southern Alaska. The expulsion of sediment-laden fluids from below ground could form a large subsurface cavity that causes the overlying layers to subside. The New Madrid earthquakes caused subsurface water and sand

Figure 59 Reelfoot Lake, Tennessee, created by flooding of downdropped crust from the 1811–1812 New Madrid, Missouri, earthquake. Courtesy of USGS

to spout to the surface, leaving void spaces in the ground that caused compaction of subterranean materials and subsidence.

Sediments also subside significantly when water is added to them. This condition is especially prevalent in the heavily irrigated dry western states. The land surface has lowered 3 to 6 feet on average and as much as 15 feet in the most extreme cases. The settling occurs when dry surface or subsurface deposits are extensively wetted for the first time since their deposition following the last ice age. The wetting causes a reduction in the cohesion between sediment grains, allowing them to

Figure 60   Mount St. Helens following the May 18, 1980, eruption, which blew off the upper peak.
Photo by MSH-Brugman, courtesy of USGS

move and fill in intergranular openings. The compaction produces an uneven land surface, resulting in depressions, cracks, and wavy surfaces.

RESURGENT CALDERAS

If a volcano decapitates itself by blowing off its upper peak (Fig. 60) or collapses into a partially emptied magma chamber, it forms a caldera, a Spanish word meaning "cauldron." Indeed, such a cataclysmic collapse is what keeps volcanoes from becoming the tallest mountains. A resurgent caldera forms when the sudden ejection of large volumes of molten rock from a magma chamber lying just a few miles beneath the surface abruptly removes the underpinning of the chamber's roof, causing it to collapse, leaving a deep, broad depression on the surface (Fig. 61). The infusion of new molten rock into the magma chamber slowly heaves the caldera floor upwards, causing a vertical uplift of several hundred feet. Generally, if a large part of the caldera floor bulges rapidly at a rate of several feet a day, a major eruption follows within a few days.

Resurgent calderas usually develop over a mantle plume, or hot spot, that melts near-surface rocks. They are recognized by widespread secondary volcanic activity, such as hot springs and geysers. Yellowstone is such a caldera. Like all resurgent calderas, the Yellowstone Caldera formed above a mantle plume that was large and long-lasting enough to melt huge volumes of rock. About 600,000 years ago, a massive volcanic eruption ejected some 250 cubic miles of ash and pumice, creating the Yellowstone Caldera. The caldera's floor has slowly domed upward on average about three-quarters of an inch per year since 1923.

Many other calderas, no more than a few tens of millions of years old, lie in a broad belt, covering Nevada, Arizona, Utah, and New Mexico. A million years ago, a massive eruption created the Valles Caldera in northern New Mexico, which has been explored for its geothermal energy potential. The calderas generally exist where the crust is thinning and the mantle rises close to the surface. Such geologically young calderas are inherently unstable, spurring episodes of unrest not associated with eruption. Generally, caldera eruptions are more the excep-

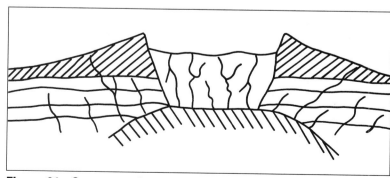

Figure 61   Structure of a resurgent caldera from the collapse of a volcano.

tion than the rule. Nevertheless, calderas are dynamic and in a delicate equilibrium, so that even small disturbances can lead to instability.

The Long Valley Caldera east of Yosemite National Park, California, was formed by a cataclysmic eruption 700,000 years ago that resulted in a 20-mile-long, 10-mile-wide, and 2-mile-deep depression (Figs. 62A & 62B). The eruption fragmented the nearby mountains into rocky debris, and some 140 cubic miles of material were strewn over a wide area as far as the East Coast.

Figures 62A & 62B   Long Valley Caldera, California. Courtesy of USGS

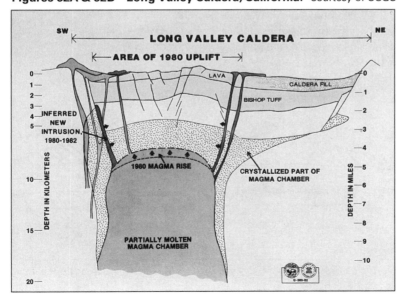

Magma appears once again to be moving into the resurgent caldera from a depth of several miles beneath the surface. The increase in volcanic and seismic activity is indicated by a rise in the center of the caldera's floor of a foot or more since 1980. A number of medium-size earthquakes of magnitude 6 or less have struck the region over the same period. These quakes might indicate that magma is pressing toward the surface and that the caldera is poised for its first eruption in 40,000 years. If an eruption does occur, large portions of neighboring Nevada could be flooded with thick basalt flows.

Similar eruptions have taken place in other parts of the world within the last million years. In northern Sumatra, the giant Toba Caldera, which is nearly 60 miles wide at its maximum, is the largest resurgent caldera on Earth. It formed 75,000 years ago, when a massive eruption, possibly the largest in the last million years, caused the crust to collapse as much as a mile or more. The caldera is presently filled with a large lake, wherein lies a 25-mile-long by 10-mile-wide island that was formed by the caldera's upraised floor.

Other types of calderas form in areas where the crust has been fractured, allowing magma to move toward the surface. The magma intrusion domes the overlying crust upward, creating a shallow magma chamber that contains a large volume of molten rock. The doming process

Figure 63 Crater Lake, Klamath County, Oregon, created by the collapse of Mount Mazama. Courtesy of USGS

produces stresses in the surface rock that forms the roof of the magma chamber, causing it to weaken and collapse along a ring fracture zone. This becomes the outer wall of the caldera after the eruption. If the caldera fills with water, it forms a deep lake similar to Oregon's Crater Lake (Fig. 63), caused by the collapse of the Mount Mazama Volcano 6,000 years ago.

COLLAPSE STRUCTURES

Limestone and other soluble rocks underlie large portions of the world. When acidic groundwater percolates through these rocks, it dissolves soluble minerals such as calcite, forming cavities or caverns. Rainwater filtering through overlying sedimentary layers reacts with carbon dioxide to form a weak carbonic acid, which flows downward through cracks in the lower rock layers and dissolves calcite or dolomite. This action enlarges fissures in the rock and eventually creates a path for more acidic water.

Land overlying these caverns can suddenly collapse, forming sinkholes of 100 feet or more in depth and up to several hundred feet across. One of the most dramatic examples of this phenomenon occurred when a sinkhole 350 feet wide and 125 feet deep suddenly opened under Winter Park, Florida, in May 1981, collapsing part of the town. At other times, the land surface can settle slowly and irregularly. The subsidence can cause extensive damage to buildings and other structures located over the pits formed by dissolving soluble minerals. Although the formation of sinkholes is a natural phenomenon, the process can be accelerated by the withdrawal of groundwater or the disposal of water into the ground. Often, sinkholes fill with water to form small lakes.

The landscape formed by numerous sinkholes is called karst terrain. Throughout the world, about 15 percent of the land surface rests on such terrain, creating millions of sinkholes. The major locations of karst terrain and caverns in the United States are in the Southeast and Midwest, as well as portions of the Northeast and West. In Alabama, soluble limestone and other sediments cover nearly half the state, and thousands of sinkholes pose serious problems for highways and other construction projects. A third of Florida is underlain by eroded limestone at shallow depths, which is subject to the formation of sinkholes that are often filled with water (Fig. 64).

Blue holes are sinkholes submerged by the sea and appear dark blue because of their great depth. Many blue holes dot the shallow waters surrounding the Bahama Islands southeast of Florida. They formed during the last ice age, when the ocean dropped several hundred feet, exposing the seabed well above sea level. The sea lowered in response to the growing ice sheets that covered the northern regions of the world, locking up huge quantities of water.

During the seabed's exposure as dry land, acid rainwater seeping into the soil dissolved the limestone bedrock, creating vast subterranean caverns. Under the weight of the overlying rocks, the roofs of the caverns collapsed, forming huge, gaping pits. At the end of the ice age, when the ice sheets melted, the area was inundated by the sea, which submerged the sinkholes. Blue holes can be very treacherous places because they often have strong eddy currents or whirlpools that can suck an unwary boat to the bottom.

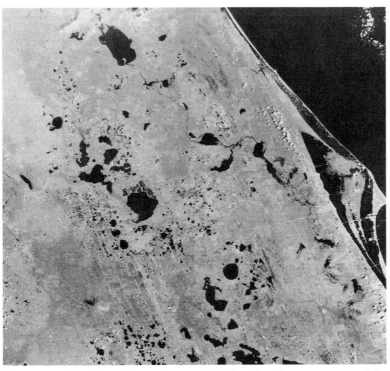

Figure 64   The Cape Kennedy area, Florida, showing numerous sinkholes filled with water. Courtesy of NASA

Figure 65 A collapsed lava tunnel, Craters of the Moon National Monument, Idaho. Photo by H. T. Sterns, courtesy of USGS

Shallow tunnels in ancient lava flows can collapse, another cause of ground subsidence. Lava tunnels are long caverns beneath the surface of a lava flow, created by the withdrawal of lava as the surface cools and hardens. In exceptional cases, they can extend up to 12 miles inside a lava flow. Often, circular or elliptical depressions exist on the surface of lava flows due to the collapse of lava tunnel roofs (Fig. 65). This phenomenon is common in the volcanic fields of Alaska, Washington, Oregon, California, and Hawaii. Among the largest is a collapse depression in a lava flow in New Mexico that is nearly a mile long and 300 feet wide.

The collapse of abandoned underground coal mines, especially those in the eastern United States, might leave the rocks above the mine workings with inadequate support and cause the surface to drop several feet, forming several depressions and pits. In situ coal gasification and oil shale retorting also can cause the overlying ground to subside. Solution mining, using large volumes of water pumped into the ground to remove soluble minerals, such as salt, gypsum, and potash, produces huge underground cavities that can collapse and cause surface subsidence. If the mines exist under towns, the overlying buildings might be heavily damaged or destroyed.

6

FLOODS

Floods are naturally occurring events that become hazardous only when people build in flood-prone areas. Floodplains are the border regions of rivers and a valuable natural resource that must be managed properly to prevent flood damage. Unfortunately, improper use of floodplains has led to the destruction of property and the loss of lives (Table 3 & Fig. 66). Too often it takes a flood to make people finally recognize the importance of proper floodplain management.

TABLE 3   THE MOST DESTRUCTIVE U.S. FLOODS

Date	Rivers or Basins	Damage (million $)	Death Toll
1903	Kansas, Missouri, and Mississippi	40	100
1913	Ohio	150	470
1913	Texas	10	180
1921	Arkansas	25	120
1921	Texas	20	220
1927	Mississippi	280	300

FLOODS

Date	Rivers or Basins	Damage (million $)	Death Toll
1935	Republican and Kansas	20	110
1936	Northeast U.S.	270	110
1937	Ohio and Mississippi	420	140
1938	New England	40	600
1943	Ohio, Mississippi, and Arkansas	170	60
1948	Columbia	100	75
1951	Kansas and Missouri	900	60
1952	Red	200	10
1955	Northeast U.S.	700	200
1955	Pacific Coast	150	60
1957	Central U.S.	100	20
1964	Pacific Coast	400	40
1965	Mississippi, Missouri, and Red	180	20
1965	South Platte	400	20
1968	New Jersey	160	—
1969	California	400	20
1969	Midwest	150	—
1969	James	120	150
1971	New Jersey and Pennsylvania	140	—
1972	Black Hills, S. Dakota	160	240
1972	Eastern U.S.	4,000	100
1973	Mississippi	1,150	30
1975	Red	270	—
1975	New York and Pennsylvania	300	10
1976	Big Thompson Canyon	—	140
1977	Kentucky	400	20
1977	Johnstown, Penn.	200	75
1978	Los Angeles	100	20
1978	Pearl	1,000	15
1979	Texas	1,250	—
1980	Arizona and California	500	40
1980	Cowlitz, Wash.	2,000	—
1982	Southern California	500	—

Date	Rivers or Basins	Damage (million $)	Death Toll
1982	Utah	300	—
1983	Southeast U.S.	600	20
1993	Midwest U.S.	12,000	24

HAZARDOUS FLOODS

The greatest dam disaster in North America took place on the south fork of the Little Conemaugh River, 13 miles northeast of Johnstown, Pennsylvania. An earthen dam over 70 feet high and 900 feet across, built in 1852 as part of a canal transportation project, was later abandoned in favor of

Figure 66   Severe flooding at Fairbanks, Alaska, on August 15, 1967. Photo by J. M. Childers, courtesy of USGS

the railroad, leaving the reservoir mostly unused and neglected. On May 31, 1889, heavy spring rains rapidly raised the level of the reservoir, causing the water to flow over the top of the dam and spout from its foundation. The bulging reservoir pushed aside the weakened dam, and a 40-foot-high wall of water raced down the valley below. One by one, small communities downstream were swept along by the raging waters. Just 15 minutes after the dam break, the floodwaters reached Johnstown, with a population of 30,000.

The water raced at an incredible speed along the constricted valley as the wall of water crashed down on the city. The flood obliterated everything in its path, and carried people off to their deaths. Hundreds of houses and other debris piled up in front of a sturdy stone railroad bridge that spanned the river in the center of town. The huge jumble of debris subsequently caught fire, and as many as 2,000 people trapped in the wreckage burned to death. Estimates of the number of people killed within 20 miles downstream of the dam ranged from 7,500 to as many as 15,000.

Another major dam-break flood occurred at the Teton Dam near Newdale, Idaho on June 5, 1976. As the Teton Reservoir behind the 130-foot-high earthen dam was being filled, seepages began in the dam wall, which severely eroded the downstream dam embankment. The weakened embankment subsequently fell into the reservoir, which breached the dam as water cascaded into the canyon below (Fig. 67). As the fast-moving floodwaters emerged from the canyon mouth 5 miles downstream, the flood waves spread rapidly over the widening floodplain of the Teton River.

The dam break caused a flood of unprecedented

Figure 67   The June 5, 1976, Teton Dam break, which caused extensive flooding downstream. Courtesy of USGS

Figure 68 A wrecked house and other debris near Drake, Colorado, from the July 31, 1976, Big Thompson River flood. Courtesy of USGS

magnitude on the Teton River, lower Henrys Fork, and Snake River. A wall of water up to 16 feet high devastated communities downstream of the dam. The rampaging flood waters carried off large trees and debris from destroyed buildings and other structures. The water spread over an area of more than 180 square miles, causing damages of about $400 million. Luckily, because of advance flood warnings only 11 lives were lost.

One of the most disastrous flash floods raged through the Big Thompson River Canyon east of Rocky Mountain National Park in north-central Colorado on July 31, 1976. Thunderstorms in the canyon area dumped some 10 inches of rain in a 90-minute interval. Billions of gallons of water rushed down the steep slopes overlooking the river and poured into the narrow canyon. The river rose with such furious speed it created a terrifying flood. People vacationing in the canyon fled uphill for their lives as a 20- foot-high wall of water bore down on them.

For almost the entire 25-mile stretch of the canyon, people scampered for the safety of the high ground ahead of the sudden onslaught of water.

The floodwaters carried off buildings, vehicles, and large trees (Fig. 68). The flood destroyed the town of Drake between Estes Park and Loveland and washed out almost the entire length of the highway through the canyon, leaving thousands of people stranded. The flood also wiped out several small communities, resulting in over $35 million in damages. At least 139 lives were lost and hundreds of people were injured.

The 1980 eruption of Mount St. Helens in southwestern Washington produced major debris flows and mudflows and severe flooding on the Toutle and lower Cowlitz rivers. Heavy runoff from melted glaciers and snowfields on the volcano's flanks supplemented by outflows from Spirit Lake below the volcano were the main sources of the floodwaters. Great volumes of sediment and thousands of fallen trees transported during the flood on the Toutle River destroyed most of its bridges (Fig. 69). Much of the sediment carried downstream into the Cowlitz and Columbia rivers formed a shoal that blocked shipping for several days.

The deadliest natural disaster in American history occurred in the Galveston, Texas, area on September 8, 1900. A hurricane storm surge swept through town, crumbling buildings and sending people into the surging coastal waters. When calm finally returned, 10,000 to 12,000 people were found dead. Similarly, the crowded Bay of Bengal, Bangladesh, is frequently pounded by cyclones on the Indian Ocean. On May 24, 1985, a powerful cyclone surged up the bay, accompanied by 15- to 50-foot-tall waves. When the storm was done, upwards of 100,000 people were dead and 250,000 others were homeless.

During the winter of 1982–1983, powerful storms brought destructive winds, tides, floods, and landslides to the California coast that caused more than $300 million in damages and forced 10,000 people to evacuate their homes. As the storms marched eastward, they drenched the southwestern states and placed many parts of America's southland underwater, forcing the evacuation of thousands of people from their homes. A huge snowpack in the Colorado Rockies brought the Colorado

Figure 69   Destruction of the St. Helens bridge caused by a mudflow from the 1980 Mount St. Helens eruption. Photo by MSH-Schuster, courtesy of USGS

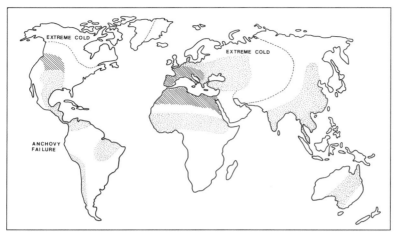

Figure 70   Strange weather of 1972 caused by El Niño. Stippled areas were affected by extreme drought. Hashed areas were affected by unusually wet weather.

River to flood stage. The unusual weather was blamed on an El Niño event, which is a warming of the eastern Pacific Ocean that occurs every five to eight years (Fig. 70). In terms of economic costs and human misery, the string of phenomenal storms could go down in the record books as possibly the worst weather event of the century.

Another strong El Niño event might have been responsible for the Midwest floods in the United States during the spring and summer of 1993. The upper tropospheric jet stream was stationary over the upper Midwest, where it steered strong weather systems into the region. Major rivers like the Mississippi and Missouri overran their levees and drowned adjacent floodplains, leaving tens of thousands of people homeless and destroying millions of acres of cropland.

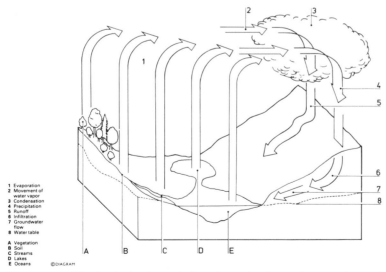

1 Evaporation
2 Movement of
   water vapor
3 Condensation
4 Precipitation
5 Runoff
6 Infiltration
7 Groundwater
   flow
8 Water table

A Vegetation
B Soil
C Streams
D Lakes
E Oceans

©DIAGRAM

It was the costliest flood in the nation's history, amounting to about $12 billion in damages. The flooding is considered a man-made disaster. Typically, levees constructed to protect property severely restrict river flow during flood stages, and reservoirs built to contain normal floods tend to overflow during massive flooding. Furthermore, upstream floodplains and wetlands act as sponges, soaking up excess flood waters; levees restrict this function, causing serious flooding downstream.

Figure 71   The hydrologic cycle involves the flow of water from the ocean onto the land and back into the sea.

THE HYDROLOGIC CYCLE

The movement of water from the ocean, over the land, and back to the sea again is known as the hydrologic cycle (Fig. 71), one of nature's most important cycles. Without the flow of water over the Earth, life as we know it could not exist. The oceans cover over 70 percent of the planet's surface to an average depth of more than 2 miles. The total volume of seawater is nearly a quarter billion cubic miles. Every day, one trillion tons of water rains down on the surface, mostly over the sea.

Water travels from the ocean to the atmosphere, crosses the land, and empties into the sea, taking on average about 10 days to complete the trip. The journey is only a few hours long in the tropical coastal areas and up to 10,000 years in the polar regions. Snow falling on the polar ice sheets reenters the sea as icebergs when glaciers plunge into the ocean. The quickest route water takes to the ocean is by runoff from streams and rivers. This is the most apparent as well as the most important part of the hydrologic cycle. Rivers provide waterways for commerce and water for hydroelectric power, municipal water supplies, recreation, and irrigation (Fig. 72). For this reason, many of the world's major cities are situated near waterways.

The total amount of precipitation the land receives is about 25,000 cubic miles annually. Some 10,000 cubic miles is surplus water that is lost by floods, held by soils, or

Figure 72 River water irrigation near Grand Junction, Colorado.

contained by wetlands. About a third of the total is base flow, which is the stable runoff of all the world's rivers and streams. The rest is groundwater flow. Some 15,000 cubic miles of water evaporate from lakes, rivers, aquifers, soils, and plants per year.

Much of the water the land receives is lost by floods, which are important for the distribution of soils over the land. During a flood, a river might change its course many times on its journey to the sea. Surface runoff supplies minerals and nutrients to the ocean and cleanses the land. The importance of water to life is so obvious it is too often overlooked. Much of the surface and subsurface water has become polluted by human activities, and the accumulation of toxic substances in the ocean from polluted runoff threatens to cause irreparable damage to marine life.

HYDROLOGIC MAPPING

Hydrologic mapping using satellite imagery can provide useful data on snow cover, sea ice extent, river flow, and flood inundation. The 1973 and the 1993 Mississippi River floods, the 1978 Kentucky River flood, and the 1978 Red River of the North flood resulted in near record-breaking water levels that were identified on satellite imagery. Regional snow cover maps are important for predicting the amount of runoff during spring thaw. The unusually stormy winter of 1982–1983 produced a larger than normal snowpack in the Rocky Mountains of Colorado, and during the spring thaw the Colorado River reached record flood stage. A repeat performance occurred in the spring of 1993.

Images obtained from weather satellites are used to monitor river basin snow cover in the United States and Canada. Several government agencies and private concerns such as utility companies use the river basin snow maps. The snow data also aid in dam and reservoir operations as well as in the calibration of runoff models. These models are designed to simulate and forecast daily streamflow in basins where snowmelt is a major contributor to runoff. This is particularly important in the West for preparing seasonal water supply forecasts.

Several methods are used to analyze satellite data for mapping snow cover. The simplest uses an optical transfer device to magnify and rectify the satellite imagery so it overlies a standard hydrologic basin map. The snow line is then manually transferred from the image to the map. Another method uses a computer to display the image data on a video monitor, and the snow line is electronically traced onto the image. A third method uses a computer to determine the snow cover pixel by pixel (tiny picture elements) by analyzing the terrain type and solar incidence angles, which control the amount of sunlight reflected back to space.

Figure 73   An ice jam in the Passumpsic River, causing flooding at St. Johnsbury Center, Vermont.
Courtesy of USDA–Soil Conservation Service.

The snow maps display the areal extent of continental snow cover but do not indicate the snow depth, which must be obtained manually in the field. The snow-cover charts are digitized and stored on computer tapes, from which are created monthly anomaly, frequency, and climatological snow-cover maps. In addition, continental or regional snow cover can be calculated over long periods for North American winter snow cover.

Satellite data are useful for detecting and locating ice cover and ice dams on rivers, especially northern rivers, where ice is particularly troublesome. Observation of river ice is important because it creates problems for hydroelectric dams, bridges, and maritime navigation. The ice becomes particularly hazardous when it breaks up and forms a dam, posing a flood threat to nearby communities (Fig. 73). Often, the ice persists because of river dams, sharp bends in the river course, or branching of the main channel by islands.

The data are also used for monitoring flash floods from large storm systems. Satellite-derived precipitation estimates and trends aid meteorologists and hydrologists in evaluating heavy precipitation events and in providing timely warnings to affected areas. Flood damage in the United States often exceeds $1 billion annually, despite the construction of flood-prevention projects to help save lives and reduce property losses. To minimize flood-related hazards, engineers and government officials need accurate information on the location of flood-hazard areas and assessments of areas of inundation when floods occur. Computer models are used to provide quick approximations of the total extent of a flood for disaster and relief planning.

FLOOD-PRONE AREAS

Floods are typically man-made disasters because people build on floodplains without recognizing the flood danger. Floodplain zoning laws and flood control projects are based on statistical analyses of relatively short-term historical records of large floods, sometimes referred to as 100-year floods. This makes assessing the risk of large floods very difficult. Moreover, due to variability in the weather, two or more record-breaking floods can occur in consecutive years.

The purpose of floodplains is to carry off excess water during floods. Failure to recognize this function has led to haphazard development in these areas with a consequent increase in flood dangers. Floodplains provide level ground, fertile soils, ease of access, and available water supplies. But because of economic pressures, these areas are being developed without full consideration of the flood risk. As a consequence, the federal government has assumed much of the responsibility for providing flood relief.

In spite of flood protection programs, the average annual flood hazard has been on the rise because people have been moving into flood-prone areas faster than the construction of flood protection projects. Therefore, the increased losses are the result not necessarily of larger floods, but of greater encroachment onto floodplains. As the population increases, more pressure is applied to develop flood-prone areas without taking proper precautions. Too often, people are uninformed about the flood risk when they build in flood-prone areas. When the inevitable flood strikes, they turn to the government to pay the cost of rebuilding.

Over 3 million miles of rivers and streams flow in the United States, and about 6 percent of the land is prone to flooding (Fig. 74). A large percentage of the nation's population and property is concentrated in these flood-prone areas. More than 20,000 communities have flood problems, and of these about 6,000 have populations over 2,500. Due to high population growth, modern floods are becoming more hazardous. For example, both the 1973 and 1993 Mississippi River floods and the 1978 Pearl River flood in Louisiana and Mississippi were three of the costliest floods in American history.

Floods threaten not only lives but also cause much suffering, damage property, destroy crops, and halt commerce. The average annual flood loss in the United States has increased from less than $100,000 at the beginning of this century to more than $3 billion today. If trends continue, by the year 2000, the annual flood loss is expected to top $4 billion, with about 100 lives lost in the United States annually.

Figure 74   Flood-prone regions of the United States.

The worst flood-related disaster in modern history was in 1887, when the Yellow River overflowed its levees and flooded much of northern China, drowning 7 million people in what has been labeled as "China's Sorrow." From 1947 to 1967, more than 150,000 people lost their lives to flooding in southern and southeastern Asia. In the same period, about 1,300 lives were lost in the United States. Many lives could have been saved if proper precautions had been taken in flood-prone areas.

FLOOD TYPES

Flash floods are local floods of great volume and short duration and are the most intense form of flooding. They generally result from torrential rains or cloudbursts associated with severe thunderstorms over a relatively small drainage area. Flash floods also occur after dam breaks or the sudden breakup of ice jams. An unusual type of flash flood resulted from the 1980 Mount St. Helens eruption, which produced major mudflows and flooding from melted glaciers and snow on the volcano's flanks, sending a torrent of mud and water laden with tree trunks into nearby rivers.

Flash floods can strike in almost any part of the nation, but they are especially common in the mountainous areas and desert regions of the American West. They are particularly dangerous in areas where the terrain is steep, where surface runoff rates are high, where streams flow in narrow canyons, and where severe thunderstorms are prevalent. Flash floods from violent thunderstorms produce flooding on widely dispersed streams, resulting in high flood waves. The discharges quickly reach a maximum and diminish almost as rapidly. Floodwaters frequently contain large quantities of sediment and debris that are collected as the river sweeps clean the stream channel.

Riverine floods (Fig. 75) are caused by heavy precipitation over large areas, by the melting of winter's accumulation of snow, or both. They differ from flash floods in extent and duration and occur in river systems whose

Figure 75 Extensive flooding on the Feather River, Sutter County, California. Photo by W. Hoffman, courtesy of USGS

tributaries drain large geographic areas and encompass many independent river basins. Floods on large river systems might continue for a few hours to several days. The floods are influenced by variations in the intensity and the amount and distribution of precipitation. Other factors that directly affect flood runoff are the condition of the ground, the amount of soil moisture, the vegetative cover, and the amount of urbanization, specifically pavement, which does not absorb water.

Figure 76 Hurricane storm surge damage to homes at Virginia Beach, Virginia, in March 1962. Courtesy of NOAA

The movement of floodwaters is controlled by the size of the river and the timing of flood waves from tributaries emptying into the main channel. As a flood moves down a river system, temporary storage in the channel reduces the flood peak. As tributaries enter the main channel, the river enlarges downstream. Since tributaries are different sizes and not spaced uniformly, their flood peaks reach the main channel at different times, thereby smoothing out the peaks as the flood wave moves downstream.

Tidal floods are overflows on coastal areas bordering the ocean, an estuary, or a large lake. The coastal lands, including bars, spits, and deltas, are affected by coastal currents and offer the same protection from the sea that floodplains do from rivers. Coastal flooding is primarily a result of high tides, waves from high winds, storm surges (Fig. 76), tsunamis, or any combination of these. Tidal floods are also caused by the combination of waves generated by hurricane winds and flood runoff resulting from heavy rains that accompany hurricanes.

Flooding can extend over large distances along a coastline. The duration is usually short and depends on the elevation of the tide, which usually rises and falls twice daily. If the tide is in, other forces that produce high waves can raise the maximum level of the prevailing high tide. The most severe tidal floods are caused by tidal waves generated by high winds superimposed on regular tides. Hurricanes are the primary source of extreme winds, and each year several of these storms enter the American mainland, causing a tremendous amount of damage and flooding, as well as severe beach erosion that continues to move the coastline landward.

DRAINAGE BASINS

A drainage basin comprises the entire area from which a stream and its tributaries receive water. For example, the Mississippi River and its tributaries drain an enormous section of the central United States from the Rockies to the Appalachians (Fig. 77). Furthermore, all tributaries emptying into the Mississippi have their own drainage areas, becoming parts of a larger basin. Each year, the Mississippi River dumps over a quarter billion tons of sediment into the Gulf of Mexico, which widens the Mississippi Delta and slowly builds up Louisiana and nearby states. The Gulf coastal states from East Texas to the Florida panhandle were built with sediments eroded from the interior of the continent and brought down by the Mississippi and other rivers.

Every year worldwide, about 25 billion tons of sediment are carried by stream runoff into the ocean, where it settles onto continental shelves. The sediment is produced when rocks are eroded by the action of wind, rain, and ice. Loose sediment grains are then carried downstream to the ocean. Rivers like the Amazon and Mississippi transport enormous quantities of sediment derived from the interiors of their respective continents. India's Ganges and Brahmaputra rivers carry about 40 percent of the world's total amount of sediment discharged into the ocean as erosion gradually wears down the Himalayan Mountains.

Individual streams and their valleys join into networks that display various types of drainage patterns, depending on the terrain. If the terrain is of uniform composition, the drainage pattern is dendritic, resembling the branches of a tree (Fig. 78). A trellis drainage pattern has rectangular-shaped tributaries and reflects differences in the bedrock's resistance to erosion. Rectangular drainage patterns also occur

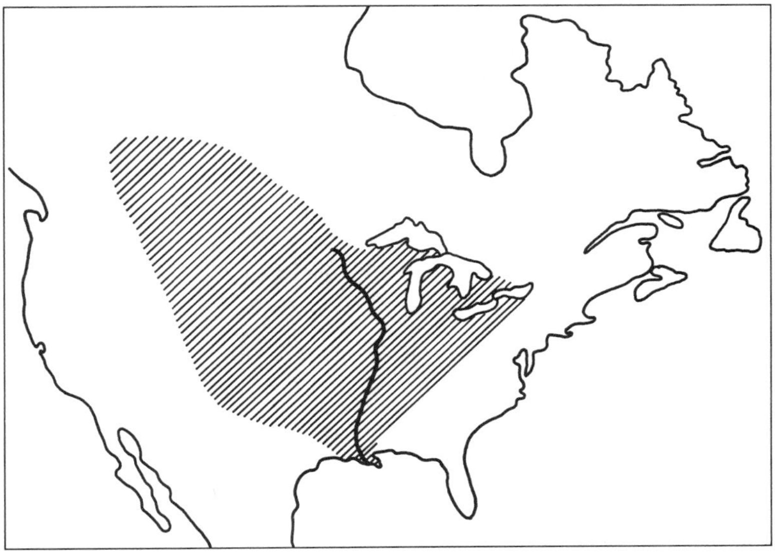

Figure 77   The drainage basin of the Mississippi River.

if the bedrock is criss-crossed by fractures, forming zones of weakness that are particularly susceptible to erosion. If streams radiate outward in all directions from a topographical high, such as a mountain, they produce a radial stream drainage pattern.

FLOOD CONTROL

The dangers of flooding can be vastly alleviated by taking certain precautions that ultimately save lives and property. The factors that control damage arising from floods include land use on the floodplain; the depth and velocity of

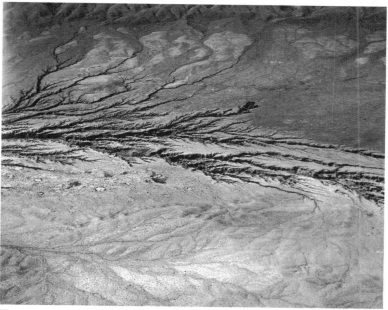

Figure 78 A dendritic drainage pattern near Green River, Utah. Photo by J. R. Balsley, courtesy of USGS

the water and the frequency of flooding; the rate of rise and duration of flooding; the time of year; the amount of sediment load deposited; and the effectiveness of storm forecasting, flood warning, and emergency services.

Direct flood effects include injury and loss of life, and damage caused by swift currents, debris, and sediment to buildings and other man-made structures. In addition, sediment erosion and deposition on the landscape might involve a considerable loss of soil and vegetation. Indirect flood effects include short-term pollution of rivers, the disruption of food supplies, the spread of disease, and the displacement of people who have lost their homes. In addition, floods might cause fires due to short circuits in high lines or breaks in gas mains.

Flood prevention involves engineering structures such as levees and flood walls that serve as barriers against high water (Fig. 79), building reservoirs to store excess water that is later released at safe rates, increasing the channel size to move water quickly off the land, and diverting channels to route flood waters around areas that require protection. The best method of minimizing flood damage in urban areas is floodplain regulation along with barriers, reservoirs, and channel improvements in flood-prone areas already developed. In response to the damage caused by the 1993 Midwest flood, the Federal government is paying home owners to move their houses to higher ground.

Municipalities should also discontinue development of areas subject to flooding that require new barriers. The most practical solution is a combination of floodplain regulations and barriers that result in less physical modification of the river system. Therefore, reasonable floodplain zoning might result in lesser utilization of flood prevention methods than if no floodplain regulations exist.

The United States has spent $10 billion on flood protection projects since the 1930s. Most of these are man-made reservoirs (Fig. 80) that even out the flow rates of rivers, with a storage capacity that can absorb increased flow during floods. The dams also generate hydroelectric power, and their reservoirs provide river navigation, irrigation, municipal water supplies, fisheries, and recreation. However, without proper soil conservation measures in the catchment areas, the accumulation of silt by erosion can severely limit the life expectancy of a reservoir.

Figure 79 Workers repair an overtopped back levee near New Orleans, Louisiana. Courtesy of Army Corps of Engineers

Floodplain regulations are designed to obtain the most beneficial use of floodplains while minimizing flood damage and cost of flood protection. The first step in floodplain regulations is flood-hazard mapping, which provides floodplain information for land use planning. The maps delineate past floods and help derive regulations for floodplain development. These controls are a compromise between the indiscriminate use of floodplains that results in the destruction of property and loss of life, as opposed to the complete abandonment of floodplains, thereby giving up a valuable natural resource. Only by recognizing the dangers of flooding and preparing for the worst can people safely utilize what nature has reserved for excess water during a flood.

Figure 80   Hoover Dam and Lake Mead on the border of Nevada and Arizona. Courtesy of USGS

7

DUST STORMS

D ust storms are awesome events that play a significant role in people's lives in many parts of the world. They can directly threaten life, and both humans and animals have died of suffocation during severe dust storms. Another direct threat dust storms pose to man is soil erosion. During the Dust Bowl years of the 1930s, the country's worst ecological disaster of the century (Fig. 81), tremendous quantities of topsoil were airlifted out of the American Great Plains and deposited elsewhere, often burying areas under thick layers of sediment. Massive dust storms raced across the prairie, carrying over 150,000 tons of sediment per square mile. Since then, agricultural practices have decreased this hazard in the United States as well as other parts of the world. Unfortunately, many regions are still at risk from soil erosion, which is seriously undermining efforts of populations to feed themselves.

DESERT REGIONS

In the last ice age, lowered precipitation levels caused desert regions to expand in many parts of the world. Desert winds blew much stronger than they do today, promoting gigantic dust storms that blocked out sunlight, furthering cooling conditions. When the great ice sheets began to retreat back to the poles, tropical regions of Africa and Arabia began to dry out during a period of rapid warming, resulting in the expansion of deserts between 14,000 and 12,500 years ago.

During an unusual wet spell from 12,000 to 6,000 years ago, many of today's African deserts were lush with vegetation and contained sev-

Figure 81   Buried machinery on a farm in Gregory County, South Dakota, during the 1930s Dust Bowl. Courtesy of USDA–Soil Conservation Service

eral large lakes. Lake Chad on the southern border of the Sahara Desert swelled to several times its present size. The Mojave and nearby deserts of the American Southwest received sufficient rainfall to sustain woodlands. Utah's Great Salt Lake expanded well beyond its present shores to occupy the adjacent salt flats. The Climatic Optimum, which began about 6,000 years ago, was a period of unusually warm, wet conditions that lasted 2,000 years. Then around 4,000 years ago, temperatures dropped significantly and the world became dryer, spawning the return of deserts.

A number of clues provide researchers with information on the climates of the past. Tree rings (Fig. 82) provide an ideal indicator of past climates, and generally the wider the rings the more favorable the climate. During a drought or a cold climate, tree rings are usually narrower due to poor growing conditions. By analyzing tree rings of the bristlecone pine, one of the longest-living plants, scientists established a drought index for the western United States, dating to the year 1600. By measuring tree rings of ancient, well-preserved trees, investigators can delve into the climate history extending back more than 7,000 years.

Deserts are the hottest and driest regions and among the most barren environments on Earth. About a third of the land surface is desert (Fig. 83). Due to human activities and natural processes, more and more land is becoming desertified, amounting to about 15,000 additional square miles a year, a little less than the size of California's Mojave Desert. Only the hardiest plant and animal species live in desert regions, including plants whose seeds can survive a 50-year drought and rodents that spend their entire lives without taking a single drink of water. Instead, they survive off water generated solely by their body's metabolism.

Deserts typically receive less than 10 inches of rainfall annually, and evaporation generally exceeds precipitation. Because of these stringent conditions, desert areas cannot support significant human populations without artificial water supplies. Much of the world's desert wasteland receives only minor quantities of rain during certain seasons, while some regions like Egypt's Western Desert, have gone essentially without rain for many years. Often, when the rains do come, they cause severe flash floods. Water levels in dry wadis rise rapidly and fall almost as fast, as the flood wave flows through the desert. Eventually, the flood waters either empty into shallow lakes that later dry up, or soak into the dry, parched ground.

Figure 82 A tree sample being prepared for annual growth ring studies. Photo by L. E. Jackson, Jr., courtesy of USGS

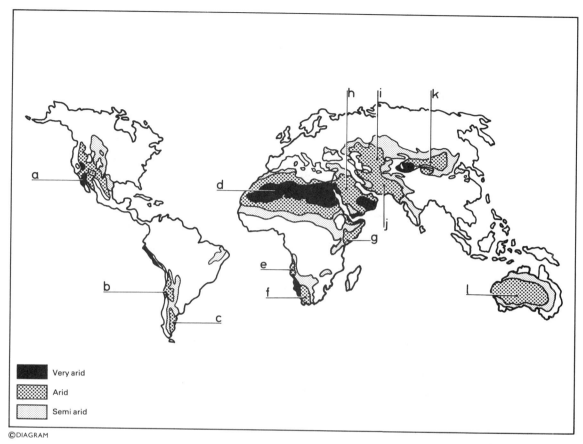

Figure 83   The world's major deserts: (a) North American, (b) Peruvian-Atacama, (c) Patagonian, (d) Sahara, (e) Namib, (f) Kalahari, (g) Somali, (h) Arabian, (i) Turkestan, (j) Iranian, (k) Gobi, (l) Great Australian.

©DIAGRAM

Most of the world's deserts lie in the subtropics north and south of the equator, in two broad bands between 15 and 40 degrees north and south latitude. In the Northern Hemisphere, a series of deserts stretches from the west coast of North Africa through the Arabian Peninsula and Iran and on into India and China. In the Southern Hemisphere, a band of deserts runs across South Africa, central Australia, and west-central South America.

After warm, moist air rises in the tropics, where precipitation levels are high, little moisture remains for the subtropics. The dry air cools and sinks, producing zones of semipermanent high pressure, called blocking highs because they block advancing weather systems from entering the region, producing mostly clear skies and calm winds. Tall mountains also tend to block weather systems by forcing rain clouds to rise, which causes them to give up their precipitation on the windward side of the range. This produces a rain shadow zone on the lee side of the range, which makes it rain deficient, spawning the growth of deserts.

Since deserts are generally light in color, they have a high albedo (Table 4), which is the ability of objects to reflect sunlight. Desert sands absorb much heat during the day, and the surface scorches at temperatures often exceeding 65 degrees Celsius (150 degrees Fahrenheit). But because the skies are generally clear at night, the thermal energy trapped in the sand quickly escapes due to the low heat capacity. This makes desert regions among the coldest environments, and even summertime temperatures can drop below freezing at night. As a result, deserts have the highest temperature extremes of any environment.

TABLE 4   ALBEDO OF VARIOUS SURFACES

Surface	Percent Reflected
Fresh-fallen snow	80–90
Old snow	45–70
Stratus clouds,	
500 feet thick	25–63
500–1000 feet thick	45–75
1000–2000 feet thick	59–84
Average all types and thicknesses	50–55
White sand	30–60
Light soil (or desert)	25–30
Concrete	17–27
Moist, plowed field	14–17
Green crops	5–25
Green meadows	5–10
Green forests	5–10
Dark soil	5–15
Blacktopped road	5–10
Water, depending on sun angle	5–60

The dry valleys of Antarctica between McMurdo Sound and the Transantarctic Range are among not only the coldest places but also possibly the most impoverished deserts on Earth. They receive less than 4 inches of snowfall each year, most of which is blown away by strong winds that can reach 200 miles per hour and more. This unique geography makes these areas ideal locations for experiments conducted for future robotic and manned excursions to Mars because the two landscapes have much in common (Fig. 84).

An interesting feature found quite by accident lies deep beneath the sands of the Sahara Desert, the largest on Earth. A search through the sands uncovered one of the last great river systems in the world that was as wide as the present-day Nile Valley. Channels hundreds of thousands of years old wind through valleys millions of years old. Dozens of human artifacts included stone axes up to 250,000 years old. They appeared to have originated from ancient campsites, where early people known as *Homo erectus* lived and made stone tools in an area that is now utterly uninhabitable. The buried valleys under the sand might once have been pathways for early humans migrating out of Africa into Europe and Asia. The discovery suggests that buried rivers might exist in other deserts as well.

Figure 84   The landscape of Mars from Viking I showing boulders surrounded by wind-blown sediment. Courtesy of NASA

DESERTIFICATION

Around 5,000 years ago, the Phoenicians migrated out of the desert of Saudi Arabia and settled along the eastern coast of the Mediterranean Sea, where they established such cities as Tyre and Sidon in what is now Lebanon. The land was mountainous and heavily forested with cedars, which became the primary source of timber for the region. When the flat plains along the coast became overpopulated, people moved to the slopes, which they cleared and cultivated, severely eroding the soil. Today, very little is left of the 1,000-square-mile forest, and the bare slopes are littered with the remains of ancient terrace walls used in a futile attempt to control erosion.

In northern Syria lie several once prosperous cities that are now dead. These ancient cities prospered by converting forests into farmland and exporting olive oil and wine. After invasions by the Persians and Arabs that destroyed the agriculture, up to 6 feet of soil eroded from the slopes. Today, after 1,300 years of neglect, the once-productive land is almost completely destroyed, a man-made desert void of soil, water, and vegetation.

The Fertile Crescent, known as the cradle of civilization, between the Tigris and Euphrates rivers in present-day Syria and Iraq, supplied food to a Middle Eastern population of 17 to 25 million people. Today, the region is mostly infertile desert due to overirrigation and soil salinization by Sumerian farmers 6,000 years ago. The world irrigates more than 10 percent of its cropland, requiring about 600 cubic miles of water annually. The United States irrigates nearly a quarter of its farmland (Figs. 85A & 85B), tripling the amount of acreage since the Second World War. Heavy use of irrigation, which not long ago turned vast stretches of America's western desert into the world's most productive agricultural land, is now ruining hundreds of thousands of acres. Perhaps we should take a lesson from the Sumerians.

The Sahel region of central Africa was once mostly tropical forest. It

Figure 85A Irrigated cotton in Imperial County, California. Courtesy of USDA–Soil Conservation Service

lies to the south of the Sahara Desert, extending in a 250-mile-wide band from coast to coast across Central Africa (Fig. 86). Over a thousand years ago, nomads of the Sahel lived by hunting and herding. They cut down and burned trees to improve grazing in the region, turning natural forests into grasslands. When colonial powers carved up Africa among themselves during the last century, they forced the people of the Sahel to settle in the region as farmers and herdsmen. Because of overgrazing and soil erosion, they turned the Sahel into an extensive man-made desert.

During the droughts of the 1970s and 1980s, the advancing sands of the Sahara Desert overran the Sahel, steadily engulfing everything in their path at a rate of 3 miles per year. In 20 years, Mauritania lost 75 percent of its grazing land to the encroaching sands of the Sahara. In the 1980s, the southern end of the desert crept 80 miles farther south. A vast belt of drought conditions now spreads across the continent to the south, parching the land and starving its inhabitants.

Desertification, which severely degrades the environment, is caused mainly by human activity and climate. Soil erosion ruins millions

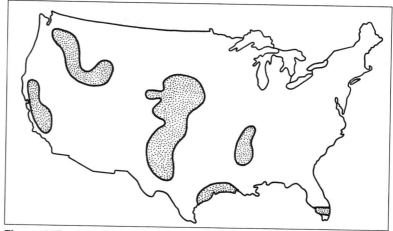

Figure 85B   Heavily irrigated areas of the United States.

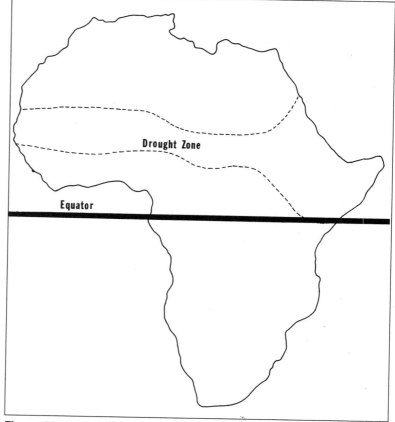

Figure 86   The Sahel region of central Africa.

109

Figure 87   The soil profile, showing organic-rich topsoil and sandy, infertile subsoil below. Measurements marked in feet. Photo by B. C. McLean, courtesy of USDA–Soil Conservation Service

of acres of once-fertile cropland and pasture every year. After the land loses its topsoil from erosion, only the coarse sands of the infertile subsoil remain (Fig. 87), thus creating a desert. The process of desertification is exacerbated because the lack of vegetation subjects the land to flash floods, higher evaporation rates, and tremendous dust storms. The denuded land also has a higher albedo, which contributes to less rainfall and further denuded land, causing man-made deserts to march across once fertile regions. Worldwide, perhaps a third or more of previously fertile land is now rendered useless by erosion and desertification. Moreover, from one-half to three-quarters of all irrigated land could be destroyed by soil salinization by the end of this century.

During the last century, global rain forests covered an area twice the size of Europe. Today, the forested area has been reduced by half for additional cropland. As developing nations attempt to raise their standards of living, they first clear the forests and drain the wetlands for agriculture. Farmers clear much of the land by wasteful slash-and-burn methods, in which trees are cut and set ablaze and their ashes used to fertilize the thin, rocky soil. A year or two of improper farming and grazing practices wears out the soil and farmers are forced to abandon their fields. The land is then subjected to severe soil erosion because plants no longer exist to protect against the effects of wind and rain. Once the soil disappears, rain forests that have been in existence for as long as 30 million years cannot return.

Countries have also cleared the tropical rain forests on an unprecedented scale for cattle grazing, which again tends to destroy the land. The beef is mostly exported to more affluent nations for relatively low prices. After a couple of years of extensive agriculture, the soil is robbed of its nutrients, and because most of the world's farmers cannot afford expensive fertilizers they are forced to abandon the infertile land. Meanwhile, heavy downpours wash away the denuded topsoil, often leaving bare bedrock behind.

The destruction of the rain forests brings changing weather patterns to the forests themselves, possibly adding to the desertification. The denuding of the world's forests increases the Earth's albedo, with a consequential loss of precipitation. Soot from massive forest fires (Fig. 88) absorbs sunlight, which

Figure 88 **The Amazon basin of South America is obscured by smoke from clearing and burning of the tropical rain forest; view from the Space Shuttle *Discovery* in December 1988.** Courtesy of NASA

Figure 89 South America's Amazon River Basin on a cloudy day during the wet season in 1973. Courtesy of NASA

heats the atmosphere and produces a temperature imbalance, causing temperatures to rise with altitude—just the opposite of what they should do. Therefore, large quantities of soot in the atmosphere generated by tropical forest fires could result in abnormal weather throughout the world.

The changing weather patterns put an additional strain on the forests and subject them to infestation and disease, causing an additional die-out of trees. Furthermore, under normal conditions high evaporation rates and transpiration rates within the forests help to create more clouds, which contribute to the prodigious amounts of rainfall these areas receive (Fig. 89). But when people remove the forests, the cycle is broken, and the torrential rains cause severe flash floods that can be disastrous for those living along streams.

Severe erosion caused by large-scale deforestation can overload rivers with sediment, causing considerable problems downstream. Monsoon floodwaters cascading down the denuded foothills of the Himalayas of northern India and carried to the Bay of Bengal by the Ganges and Brahmaputra rivers have devastated Bangladesh, where several thousand people have lost their lives to the floods. The Amazon River in South America is forced to carry much more water during the flood season due to deforestation at its headwaters. Deforestation and soil erosion are also causing many of the world's rivers to carry a higher sediment load.

HABOOBS

A severe dust storm that forms over a desert during convective instability, such as a thunderstorm, is called a haboob, an Arabic word meaning "violent wind." Dust storms arise frequently in the Sudan of northern Africa, where in the vicinity of Khartoum they are experienced about two dozen times a year. They are associated with the rainy season and remove a remarkable amount of sediment. A typical dust storm 300 to 400 miles in diameter can airlift over 100 million tons of sediment. During the height of the season, between May and October, from 12 to 15 feet of sand can pile up against any object exposed to the full fury of the storms.

Many severe dust storms occur in the American Southwest, where in Phoenix, Arizona, the frequency averages about a dozen per year. As in the Sudan, American dust storms are most frequent during the rainy season, normally July and August. Surges of moist tropical air from the Pacific rush up from the Gulf of California into Arizona and generate long, arching squall lines, with dust storms fanning out in front. These individual outflows often merge to form a solid wall of sand and dust that stretches for hundreds of miles. The sediment rises 8,000 to 14,000 feet above ground level and travels at an average speed of 30 miles per hour, with gusts of 60 miles per hour or more. Dust storms also can give rise to small, short-lived and intense whirlwinds within the storms themselves or a short distance out in front that can damage buildings and other structures in their paths. Often, people and animals die of suffocation during severe dust storms. For example, in 1895, a major dust storm in eastern Colorado is reported to have killed 20 percent of the region's cattle.

Average visibility falls to about a quarter mile but can drop to zero in very intense storms. After the storm ceases, an hour or so elapses before the skies began to clear and visibility returns to normal. However, if the parent thunderstorm arrives behind the dust storm, its precipitation clears the air in much less time. But as often happens, the trailing thunderstorm does not arrive or the precipitation evaporates before reaching the ground, a phenomenon known as virga, causing the sediment to remain suspended for several hours or even days.

Vast dust storms can result when an enormous airstream moves across the deserts of Africa. Giant dust bands 1,500 miles long and 400 miles wide often travel across the region, driven by strong cold fronts. Some large storm systems have even carried dust clear across the Atlantic to South America. The dust rises to high altitudes, where westward-flowing air currents transport it across the ocean.

Massive dust storms also arise in Arabia, central Asia, central China, and the deserts of Australia and South America, where the most obvious threat is soil erosion. Wind erosion takes more than a million acres out of production in Russia annually, making it more difficult for the country to feed itself. The deserts claim more land yearly. The tendency of the wind to erode the soil is often aggravated by improper agricultural practices. In the United States, wind erodes about 20 million tons of soil each year. The primary method of controlling wind erosion is by maintaining a surface cover of vegetation. But if rainfall is below normal such measures might not be effective and the soil simply blows away.

SAND DUNES

Wind erosion develops mainly by deflation, which is the removal of large amounts of sediment by windstorms, forming a deflation basin. In some areas, deflation produces hollows called blowouts, which are recognized by their typically concave shape. Often, after the finer material is evacuated, a layer of pebbles remains behind to protect against further deflation.

Over a period of thousands of years, deserts develop a protective shield of pebbles coated with desert varnish, composed of magnesium and iron oxides exuded from within the rocks. The pebbles become fairly well embedded and vary in size from a pea to a walnut, making it difficult for even the strongest desert winds to pick them up. This process helps to hold down the sand grains to create a stable terrain. Any disturbance on the surface can spawn a new generation of roving sand dunes and a higher incidence of dust storms that sweep the sediments from place to place.

About 10 percent of the world's deserts are composed of sand dunes, which are driven across the desert by the wind. Sand grains march across the desert floor under the influence of strong winds by a process known as saltation. The grains of sand

Figure 90   Linear dunes in the northern part of the Namib Desert, Namibia. Photo by E. D. McKee, courtesy of USGS

become airborne for a moment and upon landing they dislodge additional sand grains, which repeat the process. In this manner, sand dunes engulf everything in their path, including man-made structures, and pose a major problem in the construction and maintenance of highways and railroads that cross sandy areas of desert. Sand dune migration near desert oases poses another serious problem, especially when encroaching on villages. Methods to mitigate damage to structures from sand dunes include building windbreaks and funneling sand out of the way. Without such measures, disruption of roads, airports, agricultural settlements, and towns could become a major problem in desert regions.

The direction, strength, and variability of the wind, the moisture content of the soil, the vegetation cover, the underlying topography, and the amount of movable soil exposed to the wind determine the size and form of sand dunes. Sand dunes generally have three basic shapes, determined by the topography of the land and patterns of wind flow. Linear dunes (Fig. 90) align in roughly the direction of strong prevailing winds. Their length is substantially greater than their width, and they lie parallel to each other, sometimes with a wavy pattern. Crescent dunes, also called barchans, are symmetrically shaped with horns pointing downwind. They travel across the desert at speeds of up to 50 feet a year. Parabolic dunes form in areas where sparse vegetation anchors the side arms, while the center is blown outward, causing sand in the middle to move forward. Star dunes (Fig. 91) form by shifting winds that pile up sand into central points that can rise 1,500 feet and more, with several arms radiating outward, looking much like giant pinwheels.

A curious feature exhibited by sand dunes is an unexplained phenomenon known as booming sands. When sand slides down the lee side of a dune, it sometimes emits a loud rumbling boom, similar to the sound of a jet aircraft flying overhead. The booms can be triggered

Figure 91   Compound star dunes in Gran Desierto, Sonora, Mexico.
Photo by E. D. McKee, courtesy of USGS

by simply walking along the dune ridges. The low frequency of the boom appears to originate from a cyclic event occurring at an equally low frequency. However, normal landsliding involves a mass of randomly moving sand grains that collide with a frequency much too high to produce these mysterious noises.

DUST BOWLS

Droughts occur when precipitation activity shifts around the world (Fig. 92). In 1983, the United States experienced one of the worst droughts since the Dust Bowl of the 1930s, with crop losses approaching $10 billion. That same year Australia had the most severe drought in over 100 years. The nearly year-long drought cut grain production by about half that of the previous year. Thousands of sheep and cattle dying of starvation and thirst had to be shot and buried in mass graves.

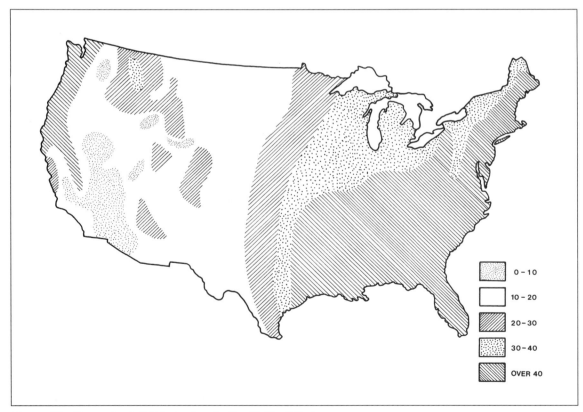

	0 – 10
	10 – 20
	20 – 30
	30 – 40
	OVER 40

Figure 92   Annual rainfall in inches in the United States.

An equally intense drought caused food shortages in southern Africa and affected West Africa and the Sahel region bordering the Sahara Desert. The sub-Saharan drought of the last quarter century was the worst in 150 years. The 1983 and 1984 droughts, which left more than a million people dead or dying of famine, were the worst of this century. In the United States, the 1980s witnessed six of the seven hottest years since the end of the Little Ice Age 140 years ago, even surpassing those of the 1930s Dust Bowl. During the 1988 drought, for the first time since World War II Americans consumed more food than they grew.

If global temperatures continue rising, the central regions of continents that normally experience occasional droughts could become permanently dry wastelands. Soils in almost all of Europe, Asia, and North America could become drier, requiring up to 50 percent additional irrigation. Rises in temperatures, increases in evaporation rates, and changes in rainfall patterns would severely limit the export of excess food for developing countries during times of famine. Dry winds of tornadic force would create gigantic dust storms and severe erosional problems.

Increasing surface temperatures could have an adverse effect on global precipitation patterns. Subtropical regions might experience a marked de-

Figure 93 Soil drifting from an unprotected corner of a wheat field in Chase County, Nebraska. Photo by H. E. Alexander, courtesy of USDA–Soil Conservation Service

crease in precipitation, encouraging the spread of deserts. Increasing the area of desert and semidesert regions would significantly affect agriculture, which would be forced to move to higher latitudes. Canada and Russia would then turn into breadbaskets, while the United States might have to import grain. Unfortunately, the soils in the northern regions are thin due to glacial erosion during the last ice age and would soon wear out from extensive agriculture. Furthermore, changing weather patterns due to instabilities in the atmosphere could make deserts out of once productive farmlands (Fig. 93).

Changes in precipitation patterns could have a profound effect on the distribution of water resources needed for irrigation. Higher temperatures would augment evaporation, causing the flow of some rivers to decline by as much as 50 percent or more, while other rivers might dry out entirely. During the 1988 drought in the United States, the Mississippi River fell to a record low, making navigation impossible over long stretches. Ancient sunken derelicts were exposed for the first time in this century, underscoring the severity of the drought.

8

GLACIERS

Glacial ice covers approximately one-tenth of the Earth's surface, and glaciers contain about 70 percent of all fresh water. Alpine glaciers are found on every continent and hold as much fresh water as all the world's rivers and lakes. If the ice caps melt during a sustained warmer climate, they could substantially raise sea levels and drown coastal regions. Beaches and barrier islands would disappear as shorelines continued to move inland (Fig. 94). Half the people of the world live in coastal areas. Rich river deltas that feed a large portion of the world's population would be inundated by the rising waters. Coastal cities would have to move farther inland or build costly seawalls to protect against the raging sea.

THE POLAR ICE CAPS

Within the last few million years, permanent ice sheets formed at both poles, a unique event in geologic history, when just having a single polar ice cap was rare. The formation of glacial ice resulted from the configuration of the Earth's crust. In our planet's early history, continents converged on areas around the equator, allowing warm ocean currents access to the polar regions, which kept them ice free year-round. During the last one

Figure 94 **Development along the shorefront of Ocean City, Maryland. The distance between these buildings and the shoreline leaves little room for natural processes during storms.** Photo by R. Dolan, courtesy of USGS

hundred million years, the continents shifted positions, so that two large land masses covered the South Pole and a nearly landlocked sea covered the North Pole.

Most of the continental landmasses moved north of the equator, leaving little land in the Southern Hemisphere, where the ocean covers nine-tenths of the surface. The drifting of the continents radically changed patterns of ocean currents, whose access to the poles was restricted. Without warm ocean currents flowing from the tropics to keep the polar regions free of ice, glaciers, which have a high albedo and thus cool the planet, could remain until such a time when the continents once again drift back toward the equator, perhaps in another 100 million years.

Greenland, the world's largest island, began rifting apart from Eurasia and North America about 60 million years ago. Occasionally, Alaska and Siberia connected and closed off the Arctic Basin, which isolated it from warm-water currents originating in the tropics, resulting in the accumulation of pack ice. About 40 million years ago, Antarctica separated from Australia and drifted over the South Pole, where it acquired a continental-size ice sheet (Fig. 95). The existence of ice at both poles established a unique equator-to-pole oceanic and atmospheric circulation system.

The Tethys Sea was a large, shallow equatorial body of water that separated Africa and Eurasia during the Mesozoic and early Cenozoic eras. Warm water, top-heavy with salt from high evaporation rates and little rainfall, sank to the bottom. Meanwhile, Antarctica, whose climate was much warmer than it is today, generated cool water that filled the upper layers of the ocean, causing the entire ocean circulation system to run backwards. About 28 million years ago, Africa collided with Eurasia, which turned off warm-water currents flowing to the poles, forming a major ice sheet in Antarctica. Cold air and ice flowing into the surrounding sea cooled the surface waters. The cold, heavy water sank to the bottom and flowed toward the equator, generating the ocean circulation system that exists today.

Some 3 percent of the planet's water is locked up in the polar ice caps, which cover on average about 7 percent of the Earth's surface. The Arctic is a sea of pack ice, whose boundary is the 10-degree-Celsius July isotherm, which is the extent of the polar drift ice during summer. A permanent ice cap did not develop over the North Pole until about 4 million years ago, when Greenland acquired its first major ice sheet. The sea ice covers an average area of about 4 million square miles, with an average thickness of 15 to 20 feet. Like ice cubes in a cold drink, its melting would not significantly raise the level of the ocean.

Figure 95   Daniell Peninsula, Antarctica. Photo by W. B. Hamilton, courtesy of USGS

By far, the greatest amount of ice lies atop Antarctica, which covers about 5.5 million square miles, an area larger than the United States, Mexico, and Central America combined. Entire mountain ranges are covered by a single sheet of ice 3 miles thick in places, with an average thickness of 1.3 miles. The total amount of Antarctic ice is approximately 7 million cubic miles, enough to make an enormous ice cube nearly 200 miles on each side.

About 37 million years ago, global temperatures plummeted due to continental movements and Antarctica accumulated a thick layer of ice that dwarfed even the present ice sheet. Sometime during the next 15 million years, most of the ice sheet melted, probably due to a warming global climate. Around 13 million years ago, a new ice sheet formed as the climate grew colder again and ocean bottom temperatures approached the freezing point. All the continent's land features, including canyons, valleys, plains, plateaus, and mountains were buried under ice, which has at times extended across the sea as far as the tip of South America. During the Pleistocene ice ages, the Antarctic ice cap expanded to become about 10 percent larger than its present size.

The snows in Antarctica accumulate into thick ice sheets because virtually no melting takes place from year to year. The summer mean monthly temperature is –35 degrees Celsius, and the winter mean monthly temperature is –60 degrees, in places dropping to –90 degrees. Barren mountain peaks soar 17,000 feet above the ice sheet, and 200-mile-per-hour winds shriek off the ice-laden mountains and high ice plateaus. The Transantarctic Range divides the continent into a large eastern ice mass and a smaller western lobe about the size of Greenland. The West Antarctic ice sheet, which lies mostly on the continental shelf anchored by scattered islands, apparently did not form until about 9 million years ago.

During the winter months, from June to September, nearly 8 million square miles of ocean that surrounds Antarctica is covered by sea ice, with an average thickness of less than 3 feet. Because of this great expanse of ice, Antarctica plays a more significant role in atmospheric and oceanic circulation than does the Arctic. The sea ice is punctured in various places by coastal and ocean polynyas, which are large open-water areas kept from freezing by pockets of upwelling warm water currents. The coastal polynyas are essentially sea ice factories because they expose portions of open ocean that later freeze, thereby continuing the ice manufacturing process.

The waters surrounding Antarctica are the coldest in the world. A thermal barrier produced by the circum-Antarctic current impedes the inflow of warm currents. The sea ice covering the ocean around Antarctica remains for at least 10 months of the year, during which time the continent is in near total darkness for 4 months, entirely out of the sun's rays. The water temperature throughout the year varies from 2 to 4 degrees below freezing, but due to its high salt content the seawater does not freeze. Neither do Antarctic fishes, which contain an antifreezelike substance in their bodies that keeps them alive during the cold winter months.

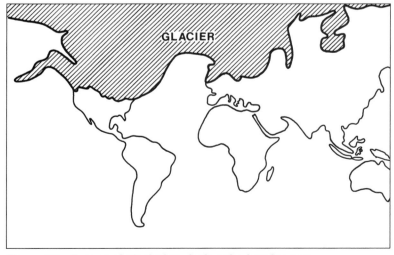

Figure 96 Extent of glaciation during the last ice age.

CONTINENTAL GLACIERS

Much of the northern regions owe their unusual landscapes to massive ice sheets that swept down from the polar regions during the last ice age (Fig. 96). Glaciers up to two miles or more thick enveloped most of upper North America and Eurasia. In places, the ice sheets scraped the crust completely clean of sediments, exposing the raw basement rock below. In other areas, the glaciers deposited huge heaps of sediment when they retreated back to the poles. Perhaps within another couple of thousand years or so, the ice sheets again will be on the move, wiping out everything in their paths. Rubble from northern cities would be bulldozed hundreds of miles to the south.

The most recent episode of glaciation began some 100,000 years ago. During its height, ice piled up as high as 2 miles or more over Canada, Greenland, and Northern Europe. North America was home to two main glacial centers. The largest ice sheet, called the Laurentide, covered 5 million square miles and spread out from Hudson Bay, reaching northward into the Arctic Ocean and southward into eastern Canada, New England, and much of the northern half of the midwestern United States. A smaller ice sheet, called the Cordilleran, originated in the Canadian Rockies and engulfed western Canada, and the northern and southern sections of Alaska, leaving an ice-free corridor in the center. Scattered glaciers also covered the mountainous regions of the northwestern United States.

Two major ice sheets existed in Europe as well, the largest of which, called the Fennoscandian, fanned out from northern Scandinavia and covered most of Great Britain and large parts of northern Germany, Poland, and European Russia. A smaller ice sheet, called the Alpine,

Figure 97   Extended shoreline during the height of the last ice age.

was centered in the Swiss Alps and covered parts of Austria, Italy, France, and southern Germany. In Asia, glaciers occupied the Himalayas and parts of Siberia. In the Southern Hemisphere, only Antarctica possessed a major ice sheet. Small ice sheets grew in the mountains of Australia, New Zealand, and the Andes of South America. Throughout other parts of the world, alpine glaciers existed on mountains that are presently ice free.

Some 10 million cubic miles of water were locked up in the continental ice sheets, which covered about one-third of the land surface, three times its present size. So much ice covered the continents that the oceans dropped as much as 400 feet. The lowered sea level extended shorelines seaward up to 100 miles or more (Fig. 97). The drop in sea level exposed land bridges that aided the migration of species, including humans, to other parts of the world. The land covered by glaciers also sank several tens of feet due to the great weight of the ice sheets.

Precipitation rates fell because the lowered temperatures allowed less water to evaporate from the sea. Since little melting took place during the cooler summers, only minor amounts of snowfall were required to sustain the ice sheets. The lower precipitation levels also increased the spread of deserts across many parts of the world. Desert winds blew more intensely than they do today, producing dust storms of gigantic proportions. So much dust was suspended in the atmosphere it blocked out sunlight, which significantly shaded the Earth, keeping it cooler than usual. Considering the high solar reflectance of the ice sheets, which also cooled the planet, it remains a mystery how the ice age ended.

Yet after some 90,000 years of gradual accumulation of ice, the mighty glaciers melted away in just a few thousand years, retreating upwards of 2,000 feet annually. About one-third of the ice melted between 16,000 and 12,000 years ago, when average global temperatures increased about 5 degrees Celsius to near present-day levels. Apparently, a renewal of the deep-ocean currents, which had been shut off or weakened during the ice age, helped thaw the planet from its deep freeze. Then between 11,000 and 10,000 years ago, during a period known as the Younger Dryas, the glaciers paused in mid-stride as temperatures suddenly fell again to ice age levels. Afterward, a second episode of melting ensued, which led to the present volume of ice by about 6,000 years ago.

While the ice sheets melted, massive floods raged across the land as water gushed from reservoirs trapped below the glaciers. Huge torrents of water surged along the Mississippi River toward the Gulf of Mexico, widening the channel several times its present size. While flowing under the ice, water surged in vast turbulent sheets that scoured deep grooves in the surface, forming steep ridges carved out of solid bedrock. Each flood continued until the weight of the ice sheet shut off the outlet of the reservoir. When water pressures built up again, another massive surge of meltwater spouted from beneath the glacier and rushed toward the sea.

Similar floods occur when a volcano erupts beneath a glacier. In Iceland, an eruption under a glacier in 1918 unleashed a flood of meltwater, called a glacial burst. In a matter of days, it released up to 20 times more water than the flow of the Amazon, the world's largest river. Several similar underglacier eruptions have occurred in Iceland during this century. Water gushing from such a glacier carves out an enormous ice cave. Geothermal heat beneath the ice creates a large reservoir of meltwater up to 1,000 feet deep, while a ridge of rock acts like a dam to hold back the water. When the dam suddenly breaks, the flow of water forms a channel under the ice that can extend up to 30 miles or more.

CAUSES OF GLACIATION

The ancient moraine and tillite deposits found on every continent suggest that at least four major periods of glaciation have occurred over the last 2 billion years (Table 5). Analysis of deep-sea sediments and glacial cores provides information on events taking place during the most recent period of glaciation, which began 2.4 to 2.6 million years ago, when a progression of ice sheets spanned the Northern Hemisphere. During this time, the surface waters of the ocean cooled dramatically, causing the demise of single-celled algae called diatoms when Antarctic sea ice extended northward and shaded the algae below. Without sunlight for photosynthesis, the diatoms simply vanished.

TABLE 5   CHRONOLOGY OF THE MAJOR ICE AGES

Years Ago	Event
2 billion	First major ice age
700 million	The great Precambrian ice age
230 million	The great Permian ice age
230–65 million	Interval of warm and relatively uniform climate
65 million	Climate deteriorates, poles become much colder
30 million	First major glacial episode in Antarctica
15 million	Second major glacial episode in Antarctica
4 million	Ice covers the Arctic Ocean
2 million	First glacial episode in Northern Hemisphere
1 million	First major interglacial
100,000	Most recent glacial episode
20,000–18,000	Last glacial maximum
15,000–10,000	Melting of ice sheets
10,000–present	Present interglacial

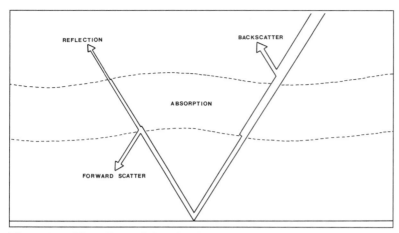

Figure 98   The effect of the albedo on incoming solar radiation.

Figure 99   The terminus of the Mer de Glace Glacier, Chamonix Valley, France–Switzerland, showing terminal and lateral moraines about the lower end of the glacier. Photo by J. A. Holmes, courtesy of USGS

A short interglacial period similar to the one we live in today followed each ice age. The last ice age began about 100,000 years ago, peaked about 20,000 years ago, and ended about 10,000 years ago. The ice sheets took a considerably longer time to reach their maximum extent than they did to recede to the comparatively insignificant amount of ice at the poles today. At the height of the ice age, glacial ice locked up 5 percent of the planet's water, which lowered sea levels several hundred feet and expanded the land area by about 8 percent.

The ice sheets formed when the climate became less seasonal and colder, but not exceptionally colder than it is today. During the last ice age, the average daytime temperatures were only about 5 degrees Celsius lower than they are at present. After the establishment of the ice sheets, they became self-perpetuating and appeared to be able to control the climate to maintain their own existence. One positive feedback mechanism by which the ice sheets sustained themselves is the albedo effect, which is the ability of objects to reflect sunlight (Fig. 98); it is mostly dependent on color. Because snow is white, it has a high reflectance, causing most sunlight to reflect out to space and not heat the surface (see Table 4).

Several triggering mechanisms could initiate an ice age. Any factor that reduces the amount of solar energy impinging on the Earth's surface and lowers global temperatures is a good candidate. The reduction of sunlight might be the result of a lower solar output. Although the sun appears to be stable today, slightly less output might have been responsible for events such as the Little Ice Age between the late 15th and mid-19th centuries, during which time sunspot activity appears to have been very low.

During this period of global cooling, average yearly temperatures fell by 1 degree Celsius, and glaciers that had been steadily retreating since the end of the last ice age suddenly began to advance. Creeping glaciers chased people out of what were once lush valleys in the northlands of Europe (Fig. 99) and froze out the Vikings, who had lived successfully on Greenland for over 500 years. When American colonists were fighting the Revolutionary War in the late 18th century, the severe winters threatened the colonials almost as much as the British army.

The Solar System has encountered enormous dust clouds in interstellar space at least a hundred times during its history. Large quantities of dust falling into the sun could disturb its normal activity. The dust also might affect the Earth's climate by forming ice clouds in the upper atmosphere, which could prevent sunlight from reaching the surface.

The Earth's orbital variations (Fig. 100) also could influence the amount of sunlight received from season to season, but they do not actually reduce the average amount of solar input during an entire year. A change in the angle of the Earth's rotational axis, called nutation, would alter the amount of solar radiation impinging on certain latitudes at different times of the year. A change in the Earth's orbit from nearly circular to elliptical would take the planet farther away from the sun during one season, therefore lowering the input of solar radiation. The precession of the spin axis changes the timing of the seasons, turning summer into winter.

Other phenomena that shut out the sun for long periods include large volcanic eruptions that loft tremendous quantities of dust and aerosols into the upper atmosphere, shad-

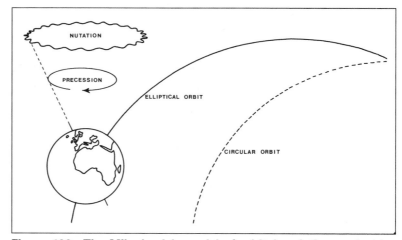

Figure 100 The Milankovich model of orbital variations coincides with ice age cycles.

ing the Earth in the process. Huge volcanic eruptions in the northern Pacific Ocean could have dropped temperatures significantly to initiate glaciation in the Northern Hemisphere. A massive meteor shower or a single large meteorite or comet impact could inject huge quantities of debris into the atmosphere and block out the sun for many years. During a geomagnetic field reversal, in which the magnetic poles reverse polarity on average every few hundred thousand years, cosmic rays heavily bombard the atmosphere, producing nitrogen oxides that could block out enough sunlight to initiate glaciation. The last magnetic reversal was about 780,000 years ago, and historically speaking the Earth is well overdue for another one.

A reduction in global atmospheric carbon dioxide, an effective greenhouse gas, also would lower global temperatures. In the last ice age, levels of carbon dioxide and methane, another potent greenhouse gas, were half their present values. The last interglacial had even higher atmospheric concentrations of carbon dioxide and was warmer than this one. The warmer seas melted the polar ice caps and caused sea levels to rise several tens of feet above their present level. Yet the warming failed to prevent the onset of the last glaciation, suggesting that today's warming trend could

Figure 101   U.S. Navy ships move a huge iceberg from a channel of broken ice leading to McMurdo Station, Antarctica. Photo by A. W. Thomas, courtesy of U.S. Navy

trigger the onset of another ice age. Indeed, according to the theory of orbital variations along with geologic evidence secured from rock cores of the seabed and from ancient coral growth patterns, the next ice age is supposedly overdue.

GLACIAL SURGE

Every year, about a trillion tons of ice discharges into the seas surrounding Antarctica and calves off to form icebergs (Fig. 101). Furthermore, the icebergs appear to be getting larger, possibly due to a warmer global climate. The number of extremely large icebergs has also increased dramatically. The largest known iceberg was located near Antarctica and measured 100 miles long, 25 miles wide, and 750 feet thick. In August 1989, it collided with the continent and broke in two.

The ice in East Antarctica is firmly anchored on land. But the ice in West Antarctica rests below the sea on bedrock and glacial till and is surrounded by floating ice that is pinned in by small islands buried below. West Antarctica is traversed by ice streams several miles broad, consisting of rivers of solid ice that flow down mountain valleys to the sea. The ice escapes through the valleys to the ice-submerged archipelago of West Antarctica, and to the great ice shelves of the Ross and Weddell seas (Fig. 102).

The banks and midsections of the ice streams might contain pools of melted water that could help lubricate the glaciers, allowing them to glide along on the valley floor and plunge into the sea. The amount of ice flowing to the coast is significantly greater than the quantity accumulating at the ice stream's source, indicating a possible instability that warrants further study, to understand the nature of ice streams and to predict their future behavior.

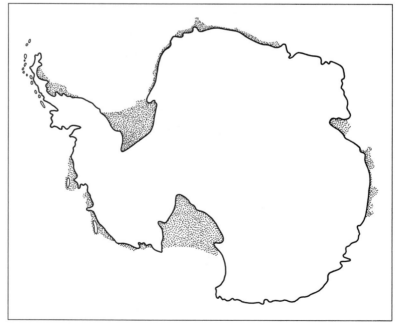

Figure 102   The Antarctic ice shelves shown in stippled areas.

The West Antarctic ice sheet could suddenly collapse due to a warmer climate. The rapidly melting, unstable ice sheet would break loose and crash into the sea. The additional sea ice would raise global sea levels upwards of 20 feet and inundate coastal areas, a hazard scientists take very seriously. Even a slow melting of both polar ice caps would raise the level of the ocean upwards of 12 feet by the end of the next century and drown much of the world's coastal plains and flood coastal cities. A rise in sea level also would lift West Antarctic ice shelves off the seafloor and set them adrift into warm equatorial waters, where they would rapidly melt and raise the sea higher still.

About 200 surge glaciers exist in North America, and some seem destined to cross the Alaskan oil pipeline. During most of their lives surge glaciers behave like normal glaciers, traveling at a snail's pace of perhaps a couple inches per day. However, at intervals of 10 to 100 years, the glaciers gallop forward up to 100 times faster than usual. The surge often progresses along a glacier like a great wave, proceeding from one section to another in a motion similar to that of a caterpillar. Subglacial streams of meltwater might act as a lubricant, allowing the glacier to flow rapidly toward the sea.

The increasing water pressure under the glacier might lift it off its bed, overcoming the friction between ice and rock, thus freeing the glacier, which rapidly slides downhill. Surge glaciers also might be influenced by the climate, volcanic heat, or earthquakes. However, many of these glaciers exist in the same areas as normal glaciers, often almost side by side. More-

Figure 103   An avalanche on Sherman glacier, Cordova District, Alaska, caused by the March 27, 1964, Alaskan earthquake. Photo by Austin Post, courtesy of USGS

over, the great 1964 Good Friday Alaskan earthquake failed to produce more surges than normal (Fig. 103).

Some 800 years ago, Alaska's Hubbard Glacier charged toward the sea, retreated, and advanced again 500 years later. Since 1895, the 70-mile-long river of ice has been flowing steadily toward the gulf of Alaska at a rate of a couple hundred feet per year. In June 1986, however, the glacier surged ahead as much as 47 feet a day. Meanwhile, a western tributary, called Valerie Glacier, advanced up to 112 feet per day. Hubbard's surge closed off Russell Fiord with a formidable ice dam, some 2,500 feet long and up to 800 feet high, whose caged waters threatened the town of Yakutat to the south.

About 20 similar glaciers around the Gulf of Alaska are heading toward the sea. If enough surge glaciers reach the ocean and raise sea levels, West Antarctic ice shelves could rise off the seafloor and become adrift. A flood of ice would then surge into the Southern Sea. With the continued rise in sea level, more ice would plunge into the ocean, causing sea levels to rise even higher, which in turn would release more ice and set in motion a vicious circle. The additional sea ice floating toward the tropics would increase the Earth's albedo and lower global temperatures, perhaps enough to initiate a new ice age. This scenario appears to have been played out at the end of the last warm interglacial, called the Sangamon, when sea ice cooled the ocean dramatically, spawning the beginning of the ice age.

RISING SEA LEVELS

At the present rate of melting, the sea could rise a foot or more by the year 2030. For every foot of sea level rise, 100 to 1,000 feet of seashore would disappear, depending on the slope of the coastline. The receding shores would result in the loss of large tracts of coastal land along with shallow barrier islands. Low-lying fertile deltas that support millions of people along with delicate wetlands, where many species of marine life hatch their young, would simply vanish (Fig. 104).

Steep waves that accompany storms at sea cause serious erosion of sand dunes and sea cliffs. The constant pounding of the surf also erodes most man-made defenses against the rising sea. Upwards of 90 percent of America's once sandy beaches are sinking beneath the waves. Barrier islands and sand bars running along the East Coast and along east Texas are disappearing at alarming rates. See cliffs are losing several feet a year to erosion, often destroying expensive homes. Most defenses used to stop beach erosion usually end in defeat as nature relentlessly batters the coast.

Over the last 100 years, the global sea level appears to have risen upwards of 6 inches due mainly to the melting of the Antarctic and Greenland ice sheets. The rapid deglaciation at the end of the last ice age between 16,000 and 6,000 years ago, when torrents of meltwater entered the ocean, raised

Figure 104 The Mosquito Lagoon, Cape Canaveral National Seashore, Florida. Courtesy of National Park Service

the sea level on a yearly basis just 10 times greater than it is rising today. The present rate of sea level rise is several times faster than it was 40 years ago, amounting to about an inch every five years. Most of the increase appears to result from melting ice sheets. The extent of polar sea ice also appears to have shrunk by as much as 6 percent during the 1970s and 1980s. Alpine glaciers, which contain substantial quantities of ice, appear to be melting as well.

During the Sangamon interglacial prior to the last ice age, the melting of the ice caps caused the sea level to rise about 60 feet higher than at present. If average global temperatures continue to rise, this interglacial could become equally as warm as the last one. The warmer climate could induce an instability in the West Antarctic ice sheet, causing it to surge into the sea. This rapid flow of ice into the ocean could raise sea levels 16 feet, inundate the continents up to 3 miles inland, and flood trillions of dollars worth of property.

9

IMPACT CRATERING

The ultimate environmental hazard is mass destruction by the impact of an asteroid or comet on the Earth. Over 100 large meteorite craters have been located throughout the world, some of which might be associated with mass extinctions of species. Occasionally, a wayward asteroid wanders near the Earth; should one ever impact the surface it would create as much havoc as a nuclear war. Indeed, the environmental consequences would be similar to those of a hypothetical nuclear winter, which would make survival difficult for all living beings.

CRATERING EVENTS

Between 4.2 and 3.8 billion years ago, thousands of large impactors the size of asteroids, bombarded the Earth and its moon (Fig. 105). All the inner planets and the moons of the outer planets display multiple craters from this massive meteorite shower, and little in the way of large-scale asteroid activity has happened since. The bombardment melted the Earth's thin basaltic crust by impact friction, and blasted apart half the crust to form huge impact basins, some with walls nearly 2 miles above the surrounding terrain and floors 10 miles deep.

During the height of the great meteorite bombardment, an enormous asteroid slammed into the North American continent in what is now central Ontario, Canada, possibly creating a crater upwards of 900 miles wide. The Earth at that time was covered by a global ocean, and the giant impact might have triggered the formation of continents. About 1.8 billion years ago, Ontario was struck again by a large meteorite, generating enough energy to melt vast quantities of rock. This impact created the Sudbury Igneous Complex (Fig. 106), the world's largest and richest nickel ore deposit. This is also the site of one of the oldest eroded impact structures, known as astroblemes. Scattered about the location are shatter cones, which are striated, conically shaped rocks fractured by shock waves generated by large meteorite impacts.

Figure 105   A large meteorite crater on the lunar surface. Courtesy of NASA

More than a mile below the floor of Lake Huron lies a 30-mile-wide rimmed circular remnant of an apparent impact structure produced by a large meteorite 500 million years ago. A crater this size would have required the impact of a meteorite about 3 miles wide. It is only one of about 130 large impact craters scattered around the world during the last 500 million years. Roughly 365 million years ago, two distinct meteorite or comet impacts on the Asian continent appear to have caused major extinctions in the late Devonian period. A large impact 210 million years ago created the 60-mile-wide Manicouagan structure in Quebec, Canada, (Fig. 107). The impact might have been responsible for the mass extinction of nearly half the existing reptile families, making room for the dinosaurs that ruled the Earth for 145 million years thereafter.

About 65 million years ago, another large meteorite supposedly struck the Earth creating a crater at least 100 miles wide whose airborne debris caused environmental chaos. Meteorite fallout material lies atop 65-million-year-old sediments worldwide. But the actual crater has yet to be found, suggesting the meteorite might have landed in the ocean. If so, it would have created a colossal tsunami that drowned coasts all around the world. A proposed site for the crater is the 110-mile-wide Chicxulub structure, the largest known meteorite crater on Earth. It lies beneath 600 feet of sedimentary rock on the northern coast of the Yucatan Peninsula. This same impact might have been responsible for the extinction of the dinosaurs along with more than half of all other species, mostly terrestrial animals and plants. Therefore, it is possible that the dinosaurs were both "created" and destroyed by large meteorite impacts.

Some 40 million years ago, two or perhaps three large meteorite impacts possibly induced another mass extinction, killing off the archaic mammals. These were large, grotesque creatures, whose disappearance paved the way for the evolution of modern mammals. In addition, major European mountain ranges were forming at this time, and the upward thrust of so much crust might have substantially cooled the planet, killing large numbers of species that could not adapt to the cold.

During the height of the last ice age, about 22,000 years ago, a large meteorite landed in northern Arizona near the present town of Winslow. The impact ejected nearly 200 million tons of rock and excavated a crater 4,000 feet wide and 600 feet deep, with a crater rim rising 150 feet above the desert floor. The massive explosion scattered pulverized rock material around the crater to a maximum depth of 75 feet. Today, the crater is a tourist attraction, where

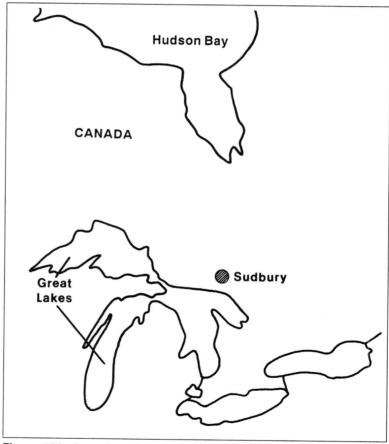

Figure 106   Location of the Sudbury Igneous Complex in Ontario, Canada.

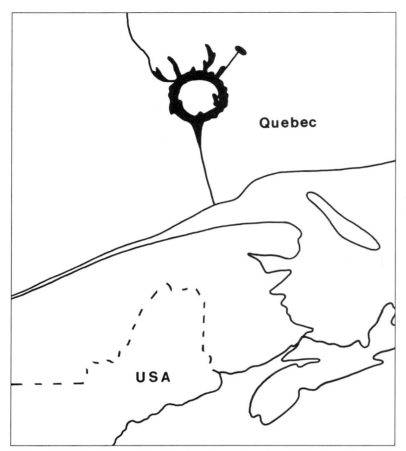

Figure 107   Location of the Manicouagan structure in Quebec, Canada.

people pay to view one of the best-preserved craters in the world. Craters formed in desert or tundra regions are better preserved because little erosion occurs in these areas.

The New Quebec Crater in Quebec, Canada, is the largest known meteorite impact structure where actual meteoritic debris has been found. It is a relatively young crater, perhaps only a few thousand years old, and has a diameter of about 11,000 feet and a depth of about 1,300 feet. The crater contains a deep lake, whose surface lies 500 feet below the crater rim. As little as 3,000 years ago, a meteorite appears to have blasted a mile-wide crater 12 miles west of Broken Bow, Nebraska. Another young impact structure, called the Wolf Creek Crater, lies near Halls Creek in Western Australia. It is a rather shallow crater, with a diameter of 2,800 feet and a depth of 140 feet. Its presence is a constant reminder that large meteorite impacts are an ongoing process, and another collision could occur at any time.

CRATERING RATES

The heavily cratered lunar highland is the most ancient region on the moon and records a period of intense bombardment around 4 billion years ago. Generally, the older the surface, the more craters that are on it. In time, the number of impacts rapidly declines, and the impact rate remains low due to a depletion of asteroids and comets. The rate of cratering appears to differ from one part of the Solar System to another. The cratering rates and the total number of craters suggest that the average rates over the past few billion years were similar for the Earth, its moon, and the rest of the inner planets.

Generally, the cratering rates for the moons of the outer planets appear to be significantly lower than those for the inner Solar System, where most of the asteroids lie. Nonetheless, the crater sizes are comparable to those on the Earth's moon and Mars (Fig. 108). Cratering rates for the moon and Mars appear to have been nearly equal. But because Mars is near a belt of asteroids that lies between it and Jupiter, the impact rates for Mars were probably higher than those for the moon.

Figure 108   Mars from Viking 1, showing numerous impact craters.
Courtesy of NASA

Major obliteration events have occurred on Mars as recently as 200 to 400 million years ago, whereas most of the scarred lunar terrain dates to billions of years ago. Mars has erosional agents, such as wind and ice, that tend to erase impact craters, while the dominant mechanism for destroying craters on the moon is other impacts. Moreover, the high degree of crater overlap makes it difficult to place the craters in their proper geologic order.

On Earth, impact craters are a few thousand to nearly 2 billion years old. Over the past 3 billion years, the cratering rate has been fairly constant, with a major impact producing a crater 30 miles or more in diameter occurring every 50 to 100 million years. As many as three large meteorite impacts that produce craters about 10 miles wide occur every million years. An asteroid half a mile in diameter, with an impact energy of a million megatons of TNT, capable of wiping out a quarter of the world's population, could strike the Earth every 100,000 years or so. A meteorite several hundred feet wide impacts every 200 to 300 years, producing the energy equivalent of a multimegaton nuclear weapon that could level a large city.

Of all the known impact craters distributed around the world, most are younger than 200 million years. The older craters are less abundant because erosion and sedimentation has destroyed them. Consequently, only 10 percent of all large craters less than 100 million years old have been discovered. Most of the known impact craters exist in stable regions in the interiors of continents because these areas experience low rates of erosion and other destructive processes.

METEORITE IMPACTS

The Earth's highly active geology has long since erased all but the faintest signs of ancient impact craters, though impacts on other bodies in the Solar System are quite evident and numerous (Fig. 109). By comparison, the Earth appears to have escaped relatively unscathed. Nonetheless, because of its larger size and greater gravitational attraction, the Earth endured several times more meteorite impacts than its next-door neighbor, the moon. The Earth was just as heavily bombarded as the rest of the Solar System, but only the vague remnants of its ancient craters remain. Many circular features appear to be impact structures, but due to their low profiles and subtle stratigraphies they previously went unrecognized as impact craters. Many geologic features once believed to be formed by forces such as uplift are now thought to be impact craters.

Figure 109 The heavily cratered terrain on Mercury. Courtesy of NASA

Dispersed throughout the world are a number of impact structures, which are large circular features created by the sudden shock of a large meteorite impacting on the surface. The structures are generally circular or slightly oval in shape and range in size from 1 to 50 miles or more wide. Some meteorite impacts form lasting, distinctive craters, while others dwindle down to only outlines of former craters. The only evidence of their existence might be a circular disturbed area that contains rocks altered by shock metamorphism. The most easily recognizable shock effect is the formation of shatter cones, caused by the fracturing of rocks into conical and striated patterns.

When a meteorite slams into the Earth's surface, it generates a tremendous shock wave with a pressure of millions of atmospheres that travels down into the rock and reflects back up into the meteorite. As the meteorite burrows into the ground, it forces the rock aside, flattening itself in the process. It is then deflected like a ricocheting bullet and ejected from the crater, followed by a spray of shock-melted meteorite and melted and

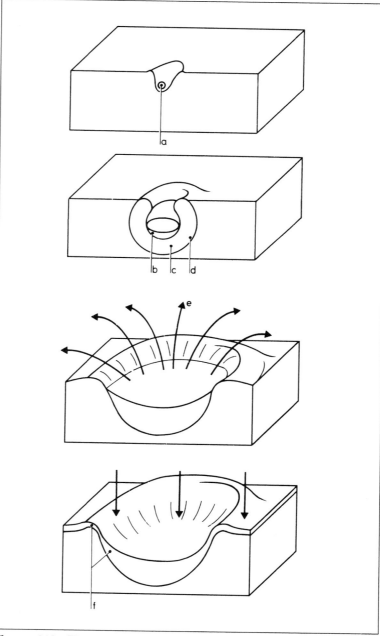

Figure 110 The formation of a meteorite crater: The meteorite punches a hole in the Earth's surface (a), forming fused rock (b), shattered rock (c), and a shock front (d). Next, large amounts of debris are ejected (e), followed by the return of fallback debris (f).

vaporized rock that shoots out at tremendous velocities. The finer material lofts high into the atmosphere, while the coarse debris falls back around the perimeter of the crater, forming a high, steep-banked rim (Fig. 110).

Large meteorites traveling at high velocities completely disintegrate upon impact, and in the process they form craters generally 20 times wider than the meteorites themselves. The crater diameter also varies with the type of rock being impacted due to the relative differences in rock strength. A crater formed in crystalline rock such as granite might be twice as large as one made in sedimentary rock. Simple craters, which are up to 2.5 miles in diameter, form deep basins. Complex craters, on the other hand, have large diameters compared to their depths and are up to 100 times wider than they are deep. The craters generally contain a central peak, surrounded by an annular trough and a fractured rim.

The Earth's highly active erosional processes have erased the vast majority of ancient impact structures. The exceptions are very large craters over 12 miles wide and more than 2.5 miles deep. These craters are so deep that even if erosion wears down the entire continent, faint remnants would still remain. Craters of extremely large meteorite impacts might temporarily reach depths of 20 miles or more and uncover the hot mantle below. The exposure of the mantle in this manner would cause a gigantic volcanic explosion, releasing more material into the atmosphere than the meteorite impact itself.

ROGUE ASTEROIDS

Asteroids (Fig. 111) are minor planets, ranging up to hundreds of miles wide. They are leftovers from the creation of the Solar System, and most form a broad band of debris that orbits the sun between Mars and Jupiter, with an inclination of about 10 degrees with respect to the ecliptic, the plane of the Solar System. Because of Jupiter's strong gravitational attraction they were unable to coalesce into a single planet. Smaller fragments, called meteoroids, are pieces broken off by constant collisions among asteroids. Due to the immense number of these fragments, meteorite falls are quite common. Daily, thousands of meteorites rain down on Earth, and occasional meteor showers can involve hundreds of thousands of stones, which almost invariably burn up on their way through the atmosphere.

Of the million or so asteroids with diameters of half a mile or more, some 18,000 have thus far been located and identified. Of these, about 5,000 have had their orbits determined precisely. The orbits of the major asteroids have been accurately plotted so that space probes headed for the outer Solar System can safely traverse the asteroid belt without a collision. Most asteroids revolve around the sun in elliptical orbits and occasionally some stretch out far enough to come within the orbits of the inner planets, including the Earth. About 60 asteroids have been observed to be out of the main asteroid belt and in Earth-crossing orbits. How they managed to fall into orbits that cross our planet's path remains a mystery. Apparently, the asteroids run in nearly circular orbits for a million or more years, then for unknown reasons their orbits suddenly stretch and become so elliptical they come within reach of our planet.

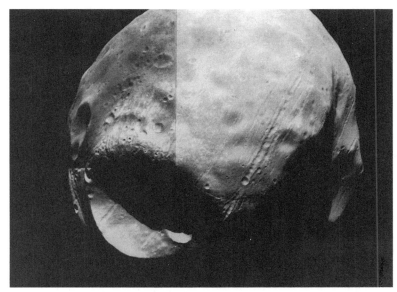

Figure 111   Phobos, the innermost moon of Mars, is thought to be an asteroid 13 miles across. Courtesy of USGS

Some Earth-crossing asteroids, called Apollo asteroids, might have begun their lives outside the Solar System and appear to be comets that have exhausted their volatile material after repeated encounters with the sun (Fig. 112). Dozens of Apollo asteroids have been identified out of a possible total of perhaps 1,000. Most are quite small and discovered only when they pass close by the Earth. Inevitable collisions with the Earth and other planets steadily depletes their numbers; their continual existence points to an ongoing source of Apollo asteroids, possibly comet nuclei.

One of the closest encounters with an asteroid occurred on October 30, 1937, when Hermes shot past the Earth at 22,000 miles per hour. The mile-

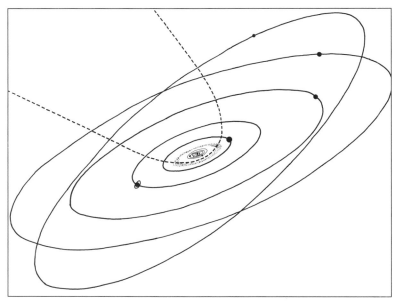

Figure 112   Comets travel in oblique orbits with respect to the Solar System.

wide asteroid missed hitting our planet by only half a million miles, or about twice the distance to the moon. In astronomical terms, that was very close. If Hermes had collided with Earth, it would have released the energy equivalent of 100,000 one-megaton hydrogen bombs. Indeed, nuclear war has many similarities to the impact of a large asteroid. The impact would send aloft huge amounts of dust and soot into the atmosphere. The debris would clog the skies and plunge the planet into a deep freeze for several months.

The closest flyby of a large asteroid in recorded history occurred on March 22, 1989, when asteroid 1989 FC came within 430,000 miles of our planet (Fig. 113). The asteroid was about half a mile wide, and though a collision with Earth would have been catastrophic, a fluke of orbital geometry might have softened the blow a little. The asteroid orbits the sun in the same direction as the Earth, completing a revolution in about one year traveling at almost the same speed as the Earth. Therefore, its approach was rather slow compared to other celestial objects. However, because of the Earth's large size, the planet's gravitational pull would have accelerated the asteroid during its final approach. If a collision had occurred, the asteroid would have produced a crater 5 to 10 miles wide, large enough to wipe out a major city.

Astronomers did not detect asteroid 1989 FC until it was already moving away from the Earth. Only then did they notice a dramatic slowdown in the asteroid's motion against background stars. To their amazement, the asteroid was rushing straight away from Earth on what must have been a near-grazing trajectory. The astronomers failed to notice the approach of the asteroid because it came from a sunward direction. Also, the moon was nearly full, further hampering observations.

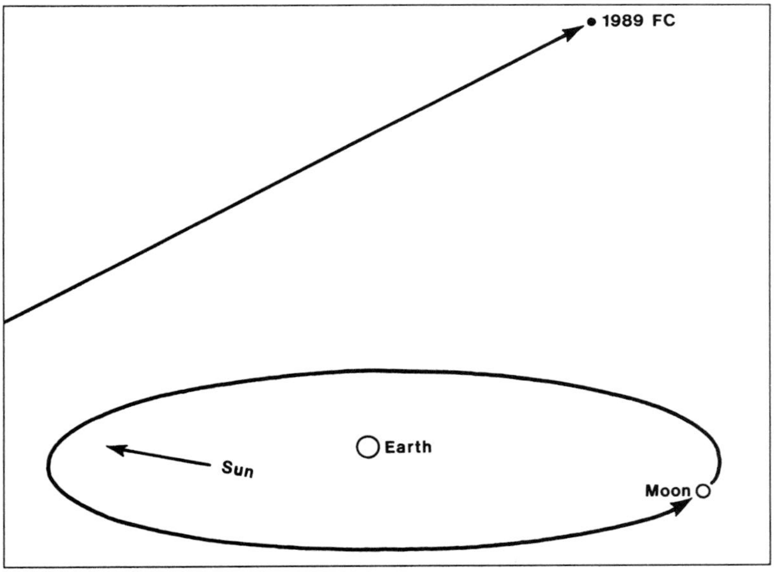

Asteroid 1989 FC is one of only 30 similar bodies that closely approach the Earth. In addition, several hundred to 1,000 or more asteroids wider than one-third of a mile are capable of crossing the Earth's orbit for a close encounter. For example, on December 8, 1992, a large asteroid called Toutatis, which

Figure 113   The closest approach of asteroid 1989 FC to the Earth.

measured 2.5 miles by 1.6 miles wide, flew within 2.2 million miles of Earth. If a collision occurred, the effects could threaten all humanity. Besides asteroids, comets have been known to fly near the Earth. The closest comet to approach the Earth was Lexell, which came within six times the distance to the Moon on July 1, 1770. The April 10, 1837, encounter with Comet Halley was close enough for the Earth's gravity to disturb the comet's orbit.

Almost all close encounters took astronomers completely by surprise, and not a single one was anticipated. To avoid the danger of an asteroid collision, the threatening body would first have to be tracked by telescopes and radar, and its course plotted accurately so its orbit could be determined precisely. If an asteroid were found to be on a collision course with the Earth, astronomers could provide timely warnings. The rogue asteroid might then be nudged out of its Earth-bound trajectory by the detonation of nuclear warheads.

STONES FROM THE SKY

Meteorite falls are a commonplace occurrence, and daily thousands of meteoroids enter the Earth's atmosphere, with occasional meteor showers that involve hundreds of thousands of stones. (The term meteoroid is used generally to refer to rocky matter in the Solar System, while meteors are meteoroids that enter and burn up in the Earth's atmosphere, and meteorites are meteors that actually survive to the Earth's surface.) Some rare meteorites found in Antarctica are thought to be rocks blasted out of Mars by large impacts. Nearly one million tons of meteoritic material are produced annually, much of which is suspended in the atmosphere where it is responsible in part for making the sky blue. Fortunately, most meteors burn up upon entering the atmosphere. The remainder that make it through the atmosphere and rain down on Earth can cause much havoc as meteorites crash into houses and automobiles.

Over 500 meteorite falls occur each year, most of which accumulate on the seafloor because the oceans cover over 70 percent of the Earth's surface. Most meteorites that fall on land are slowed down by the braking action of the atmosphere and bury themselves only a short distance into the ground. Not all meteorites are hot when they land because the lower atmosphere tends to cool them sufficiently, sometimes forming a thin layer of frost on their surfaces.

One of the oldest meteorites that remains preserved in a museum is a 120-pound stone that landed near Alsace, France, on November 16, 1492. The largest known meteorite find, called Hoba West, named for the farm it landed on, was near Grootfontein, South Africa, in 1920 and weighed about 60 tons. The largest meteorite found in the United States is the 16-ton

Willamette Meteorite, which is more than a million years old. It was discovered near Portland, Oregon, in 1902 and measured 10 feet long, 7 feet wide, and 4 feet high. One of the largest meteorites actually seen falling from the sky was an 880-pound stone that landed in a farmer's field near Paragould, Arkansas, on March 27, 1886. The heaviest observable stone meteorite landed in a cornfield in Norton County, Kansas on March 18, 1948, and dug a pit 3 feet wide and 10 feet deep.

The first recorded impact of a comet occurred in the Tunguska forest of northern Siberia on June 30, 1908 (Fig. 114). The tremendous explosion toppled and charred trees within a 20-mile radius. However, the explosion left no impact crater, suggesting a comet or stony asteroid airburst at an altitude of about 5 miles at a speed of 30,000 miles per hour. The impactor was small, estimated between 100 and 300 feet wide, which explains why no astronomical sightings were made prior to the explosion. The estimated force of the blast was as powerful as a 15-megaton hydrogen bomb. Barometric disturbances were recorded over the entire world as the shock wave circled the Earth twice. The dust generated by the explosion produced unusual sunsets and other atmospheric effects all over Europe.

The first explosion of an asteroid in modern times was observed by the pilot of a Japanese cargo plane over the Pacific Ocean about 400 miles east of Tokyo, Japan, on April 9, 1984. A cloud rapidly expanded in all directions, looking much like a nuclear detonation, except the fireball or lightning that usually accompany nuclear explosions was

Figure 114   Location of the Tunguska impact in northern Siberia.

not observed. Furthermore, an aircraft sent into the cloud to collect dust samples found no radioactivity. The mushroom cloud grew to 200 miles in diameter, rising from 14,000 to 60,000 feet in just 2 minutes. Apparently, the cloud formed by the explosion of an asteroid 80 feet in diameter, releasing the equivalent energy of a one-megaton hydrogen bomb.

If a meteor explodes as it streaks across the sky, it produces a bright fireball, called a bolide. The Great Fireball that flashed across New Mexico and nearby states on March 24, 1933, was as bright as the sun, and the meteoritic cloud grew to a towering plume in about 5 minutes. Some bolides are bright enough to be visible in broad daylight. Occasionally, their explosions can be heard on the ground and might sound like a thunderclap or the sonic boom of a jet aircraft. Everyday, thousands of bolides occur around the world, but most go completely unnoticed.

IMPACT EFFECTS

When a large extraterrestrial body, such as an asteroid or comet nucleus, slams into the Earth, the impact can produce thick dust clouds, powerful blast waves, immense tsunamis, extremely toxic gases, and strong acid rains that can cause tremendous havoc. The tsunamis generated by a splashdown in the ocean are particularly hazardous to onshore and nearshore inhabitants. Perhaps the worst environmental hazards are produced by huge volumes of suspended sediment in the atmosphere from material blasted out of the crater along with vaporized asteroidal material. Furthermore, soot from continent-sized wildfires set ablaze by the hot crater debris would clog the skies, causing darkness at noon.

If a large meteorite landed in the ocean, it would instantly evaporate massive quantities of seawater, saturating the atmosphere with billowing clouds of steam. This added burden would dramatically raise the density of the atmosphere and greatly increase its opacity, making it nearly impossible for sunlight to penetrate. Solar radiation would heat the darkened, sediment-laden layers of the atmosphere and cause a thermal imbalance that could radically alter weather patterns, turning much of the land into a barren desert. Horrendous dust storms driven by maddening winds would rage across whole continents, further clogging the skies. So much damage would beset the Earth following major impacts that subsequent mass extinctions are considered a certainty.

A large asteroid impacting on Earth could set the planet ringing like a bell. Large meteorite impacts can create so much disturbance in the Earth's crust that volcanoes and earthquakes could become active in zones of weakness. A massive impact in the Amirante Basin 300 miles northeast of Madagascar might have triggered India's great flood basalts, known as the Deccan Traps, when the subcontinent was headed toward southern Asia

65 million years ago. Quartz grains shocked by high pressures generated by a large meteorite impact found lying just beneath the immense lava flows might be linked to the impact. Such large meteorite impacts produce quartz grains with prominent striations across crystal faces. Minerals such as quarts and feldspar develop these features when high-pressure shock waves exert shearing forces on crystals, producing parallel fracture planes called lamellae.

Some geomagnetic reversals, whereby the Earth's magnetic poles switch polarities, appear to be associated with large impacts. Magnetic reversals occurring 2.0, 1.9, and 0.7 million years ago coincide with unusual cold spells. Furthermore, the last two reversals correlate with the impact of large meteorites on the Asian mainland and in the Ivory Coast region. Among the most striking examples of a large meteorite impact causing a magnetic reversal is the 15-mile-wide Ries Crater in southern Germany, which is about 14.8 million years old. A study of the magnetization of the fallback material in the crater indicates that the geomagnetic field polarity reversed soon after the impact.

A single large meteorite impact or a massive meteorite shower would eject tremendous amounts of debris into the global atmosphere, where it would block out the sun for many months or years, possibly bringing down surface temperatures significantly enough to initiate glaciation. An asteroid apparently impacted on the Pacific seafloor roughly 700 miles westward of the tip of South America about 2.3 million years ago (Fig. 115).

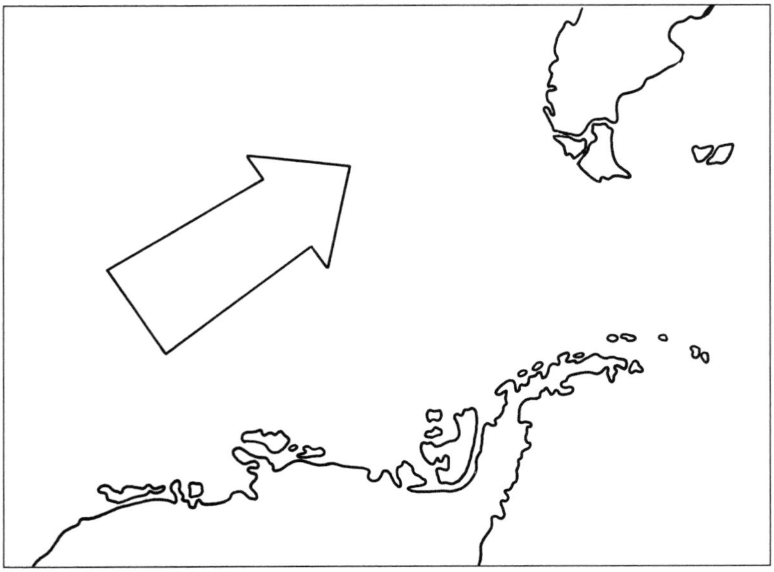

Figure 115   Point of impact of a large meteorite off the western tip of South America 2.3 million years ago.

Geologic evidence suggests that the climate changed dramatically between 2.2 and 2.5 million years ago, when glaciers began to cover large parts of the Northern Hemisphere.

If a large meteorite entered the Earth's atmosphere, air friction would produce a brilliant meteor brighter than the sun, and the searing heat would scorch everything within miles around. Shock waves generated by the meteorite's blazing speed would be strong enough to bowl people over 20 miles away. The impact would produce a rapidly expand-

Figure 116   Detonation of a hydrogen bomb, Pacific Proving Grounds. U.S. Air Force photo

ing dust plume that grows several thousand feet across at the base and extends several miles high. Most of the surrounding atmosphere would be blown away by the tremendous shock wave produced by the impact. The giant plume would turn into an enormous black dust cloud that punches through the atmosphere, like the mushroom cloud formed by the detonation of a hydrogen bomb (Fig. 116).

Heat produced by the compression of the atmosphere and impact friction that flings molten rock far and wide could set globalwide forest fires. The fires would consume some 80 percent of the terrestrial biomass, turning the planet into smoldering cinder. The impact also would send aloft some 500 billion tons of sediment into the atmosphere. A heavy blanket of dust and soot would cover the entire globe and linger

Figure 117   Regions in Europe inundated by a 300-foot tsunami created by a large meteorite impact in the North Atlantic.

for months. A year of darkness would ensue under a thick brown smog of nitrogen oxide. Waters would be poisoned by trace elements leached from the soil and rock, and acid rains would be as corrosive as battery acid.

If the asteroid landed in the ocean, upon impact it would produce a cone-shaped curtain of water, as billions of tons of seawater splashed high into the atmosphere. The atmosphere would become oversaturated with water vapor, and thick cloud banks would shroud the planet, cutting off the sun and turning day into night. The most massive tsunamis ever imagined would race outward from the impact site. The waves would traverse clear around the world, and when striking seashores they would travel hundreds of miles inland (Fig. 117) and devastate everything in their paths, making such impacts among the most destructive, if rare, geologic hazards on Earth.

10

MASS EXTINCTIONS

Considering all the great upheavals in the Earth throughout its long history, it is remarkable how life has managed to survive to the present (Fig. 118). Over 99 percent of all species that have inhabited the planet from the beginning have become extinct, so that those living today represent only a tiny fraction of the total. As many as 4 billion species of plants and animals have existed in the geologic past. Most lived during the last 600 million years, a period of phenomenal evolution of species as well as tragic episodes of mass extinctions (Table 6). Therefore, the extinction of species has been almost as prevalent as their origination. The common denominator in all mass extinctions is that biological systems were in extreme stress due to rapid and extreme changes in the environment.

TABLE 6 RADIATION AND EXTINCTION FOR MAJOR ORGANISMS

Organism	Radiation	Extinction
Marine invertebrates	Lower Paleozoic	Permian
Foraminiferans	Silurian	Permian and Triassic
Graptolites	Ordovician	Silurian and Devonian

Organism	Radiation	Extinction
Brachiopods	Ordovician	Devonian and Carboniferous
Nautiloids	Ordovician	Mississippian
Ammonoids	Devonian	Upper Cretaceous
Trilobites	Cambrian	Carboniferous and Permian
Crinoids	Ordovician	Upper Permian
Fishes	Devonian	Pennsylvanian
Land plants	Devonian	Permian
Insects	Upper Paleozoic	
Amphibians	Pennsylvanian	Permian-Triassic
Reptiles	Permian	Upper Cretaceous
Mammals	Paleocene	Pleistocene

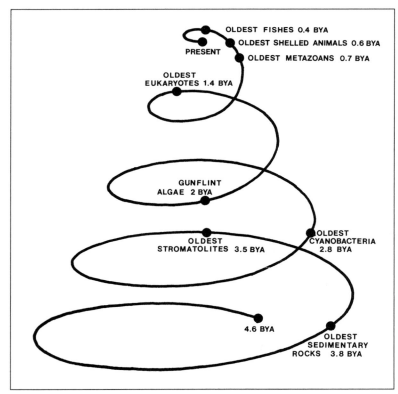

Figure 118   Geologic time spiral beginning with the formation of the Earth 4.6 billion years ago.

HISTORIC EXTINCTIONS

During the late Precambrian, when life was still sparse, great ice sheets covered more than half the land surface, instigating the first major mass extinction around 670 million years ago. The extinction decimated single-celled phytoplankton, which were the highest form of animal life at that time and the first organisms to evolve with cells containing nuclei. When the glaciation ended, a rapid population growth ensued, with a diversification of species that has no equal. By the time the Precambrian era came to a close, the seas contained large populations of widespread and diverse species. Never before or since has such a diversity of species existed, and some of the strangest animals ever known populated the planet.

The fossil record from that era is dominated by many unusual creatures, many of which probably arose by adapting to highly unstable conditions. As a consequence of overspecialization, a major extinction of species took place at the end of the Precambrian 570 million years ago. Species that survived the extinction were quite different from those left behind. These new life forms flourished in the warm climate and participated in the greatest explosion of new species in geologic history. Most of these new species, however, were not related to modern forms.

The Phanerozoic eon, from the Cambrian period beginning 570 million years ago to the present, witnessed several mass extinctions that eliminated vast arrays of species. The Cambrian was an evolutionary heyday, when the first complex animals having exoskeletons exploded onto the scene, filling the seas with a rich assortment of life. About 10 million years after the start of the Cambrian, a wave of extinctions decimated a huge variety of newly evolved species. The extinctions, which were among the most severe in Earth history, eliminated more than 80 percent of the marine animal genera. The die-offs wiped out most major groups, paving the way for the ascendancy of a famous group of invertebrates, called the trilobites (Fig. 119), which were primitive crustaceans and ancestors of today's horseshoe crab. They went on to dominate the seas for the next 100 million years.

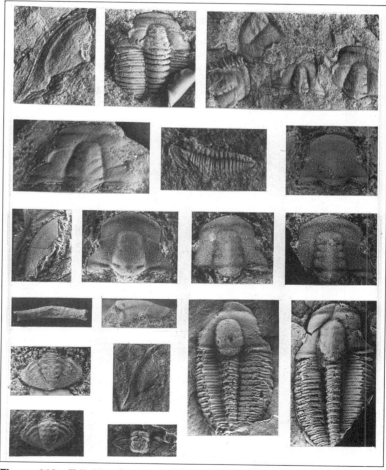

Figure 119   Trilobite fossils of the Carrara Formation in the Southern Great Basin, California–Nevada. Photo by A. R. Palmer, courtesy of USGS

A second mass extinction, which eliminated some 100 families of marine animals, occurred at the end of the Ordovician period, about 440 million years ago. During this time, glaciation had reached its peak, with ice sheets radiating outward from a center in North Africa. Most of the victims were tropical species because the tropics are more sensitive to fluctuations in the environment. Among those that went extinct were many species of trilobites. Before the extinction, trilobites accounted for about two-thirds of all species and only one-third thereafter. The graptolites, which were a strange animal species resembling a conglomeration of floating stems and leaves, also became extinct.

Another major extinction took place near the end of the Devonian period, about 370 million years ago, when many tropical marine groups simultaneously disappeared. Sponges and corals (Fig. 120), which were prolific limestone reef builders from early in the Paleozoic era, suffered an extinction from which they never fully recovered. When the corals disappeared with the receding of the seas, they were replaced by sponges and algae in the late Paleozoic. Ninety percent of the brachiopod families, which had two clamlike shells fitted face to face that opened and closed with simple muscles, also died out at the end of the Devonian.

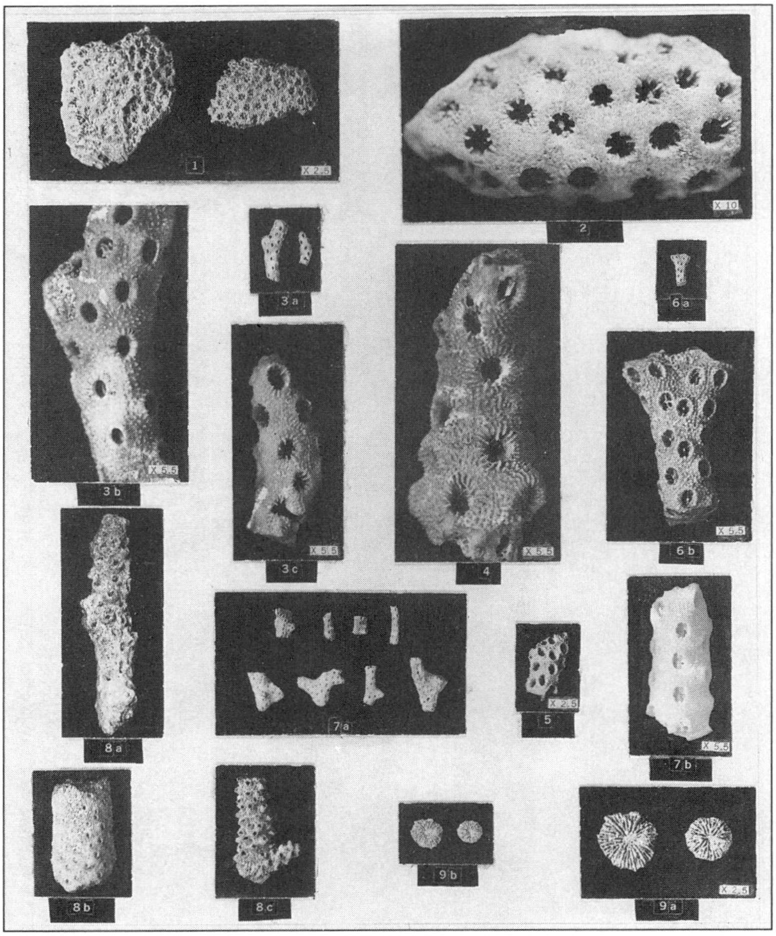

The greatest mass extinction occurred at the end of the Permian period, 240 million years ago, when half the families of marine organisms, comprising over 95 percent of all known species, abruptly disappeared. The extinction followed on the heels of a late Permian gla-

Figure 120   Fossil corals from Bikini Atoll, Marshall Islands. Photo by J. W. Wells, courtesy of USGS

ciation, and marine invertebrates that managed to escape extinction were forced to live in a narrow margin near the equator. Corals, which require warm, shallow waters, were particularly hard hit, as evidenced by the lack of coral reefs in the early part of the Triassic period. Brachiopods and crinoids, which had their golden age in the Paleozoic were relegated to minor roles during the following Mesozoic era. The trilobites, which were extremely successful during the Paleozoic, completely died out at the end of the era.

Near the end of the Triassic period, about 210 million years ago, large families of terrestrial animals began dying off in record numbers. The extinction occurred over a period of less than a million years and was responsible for killing off nearly half the reptile families. The Triassic crisis also might have eliminated tropical reef corals, possibly due to a colder climate. The extinctions forever changed the character of life on Earth and paved the way for the rise of the dinosaurs.

Many large dinosaur families, including allosaurs, brontosaurs, and stegosaurs (Figs. 121A & 121B), were cut down at the end of the Jurassic period 135 million years ago. Following the extinction, the population of small animals exploded, as species occupied niches vacated by the large dinosaurs. Most of the surviving species were aquatic creatures confined to freshwater lakes and marshes, along with small land-dwelling animals. Many of the small, non-dinosaur species were the same ones that survived the next great mass extinction, probably because of their large populations and their ability to find places to hide.

Figures 121A & 121B   Brontosaurs and stegosaurs became extinct at the end of the Jurassic period.

The most popularized extinction took place at the end of the Cretaceous period, 65 million years ago, when the dinosaurs and 70 percent of all other known species, mostly land animals, suddenly vanished. The success of the dinosaurs is exempli-

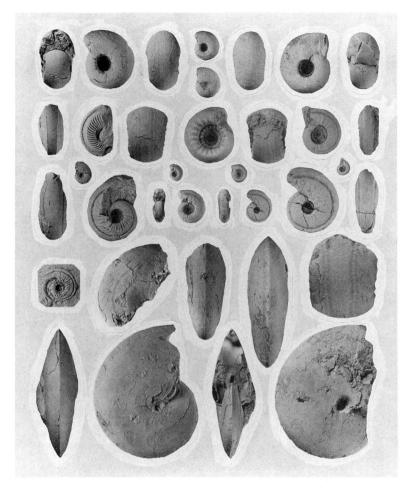

Figure 122   A variety of fossil ammonite shells. Photo by M. Gordon, Jr., courtesy of USGS

fied by their extensive range, which covered a wide variety of habitats, and their domination of all other terrestrial animals. About 500 species of dinosaurs have been classified thus far. Because the dinosaurs were not the only species to become extinct, the cause was likely external, some kind of environmental shift. Yet most mammals were not seriously affected.

The ammonites, which were cephalopods with coiled shells up to 7 feet across (Fig. 122), were fantastically successful in the warm Cretaceous seas. After surviving the critical transition from Permian to Triassic and recovering from serious setbacks in the Mesozoic, the ammonites suffered final extinction at the end of the Cretaceous, when the recession of the seas reduced their shallow-water habitats worldwide. All marine reptiles, except the smallest of turtles, became extinct at the end of the period. Ninety percent of all plankton species that lived in the surface waters of the ocean also died out.

Following the great Cretaceous extinction, life experienced an evolutionary lag, lasting up to several million years. Afterward, the mammals began to diversify rapidly, sometimes into unusual forms. A sharp extinction event occurred at the end of the Eocene epoch, 37 million years ago, when the Earth was plunged into a colder climate. The extinction wiped out most of the archaic mammals, which were large, grotesque-looking animals. Of the dozen or so orders of mammals living in the early Cenozoic era, only half lived in the preceding Cretaceous and only half are alive today.

The latest mass extinction was toward the end of the last ice age, 16,000 years ago, when giant mammals like the saber-tooth cats, ground sloths, mastodons, and woolly mammoths disappeared. When the glaciers began to retreat, a major readjustment in the global environment disrupted their food supplies, causing them to become extinct. Also, humans were becoming efficient hunters of large game and might have slaughtered some giant mammals to extinction.

CAUSES OF EXTINCTIONS

Most extinctions of the past appear to have coincided with major planetary changes, brought on by tectonic forces and lowering sea levels. Global cooling caused many extinctions as climate is possibly the most important factor influencing species diversity. As the world's oceans cool, mobile species tend to migrate to the warmer regions of the tropics. Species attached to the ocean floor and unable to move become trapped in enclosed basins and generally are the hardest hit by extinction. Only species previously adapted to cold conditions still thrive in today's oceans. Most are plant eaters that tend to be generalized feeders eating a variety of foods.

Many mass extinctions coincided with periods of glaciation because temperature is perhaps the single most important factor limiting the geographic distribution of species. Certain species, such as corals, can survive only within a narrow range of temperatures. During warm interglacial periods, species invade all latitudes. When glaciers advance across continents and oceans, temperatures drop and species are forced into warmer regions, where limited habitats exist. The intense competition for habitat and food severely limits species diversity and therefore the total number of species.

Not all climate cooling resulted in glaciation. Nor did all extinctions follow a drop in sea level caused by growing glaciers. During the Oligocene epoch, which began about 37 million years ago, seas that overrode the continents drained away as the ocean withdrew to one of its lowest levels in several hundred million years. Although the sea level remained depressed for 5 million years, little or no excess extinction of marine life occurred. Therefore, crowding conditions brought on by lowering sea levels cannot be responsible for all extinctions. Furthermore, during many mass extinctions, the sea level was not much lower than it is today.

In the final stages of the Cretaceous period, when the level of the ocean began to drop and seas were departing from the land, the temperatures in a broad tropical ocean belt known as the Tethys Sea began to fall. The change in sea level might explain why the Tethyan fauna that were the most temperature sensitive suffered the heaviest extinction at the end of the period. Species that were amazingly successful in the warm waters of the Tethys suffered total decimation when temperatures dropped. After the

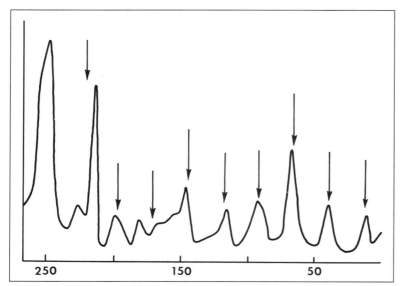

Figure 123   The 26-million-year periodicity of mass extinctions. Age in millions of years.

extinction, marine species acquired a more modern appearance as ocean bottom temperatures continued to plummet.

During the past 250 million years, 11 distinct episodes of flood basalt volcanism occurred throughout the world. Catastrophic volcanic episodes apparently took place at intervals of 200 million years, with lesser events spaced roughly 30 million years apart, corresponding to an apparent 26- to 32-million-year periodicity of extinction (Fig. 123). These were relatively short-lived events, with major phases spanning less than 3 million years. The timing of major volcanic outbreaks correlates well with the occurrence of mass extinctions of marine species (Table 7). Extensive volcanism occurring at the end of the Cretaceous might have killed off the dinosaurs and most other land species along with many marine species.

TABLE 7   FLOOD BASALT VOLCANISM AND MASS EXTINCTIONS

Volcanic Episode	Million Years Ago	Extinction Event	Million Years Ago
Columbian River, USA	17	Low-mid Miocene	14
Ethiopian	35	Upper Eocene	36
Deccan, Indian	65	Maastrichtian	65
		Cenomanian	91
Rajmahal, Indian	110	Aptian	110
South-West African	135	Tithonian	137
Antarctica	170	Bajocian	173
South African	190	Pliensbachian	191
E. North American	200	Rhaectian/Norian	211
Siberian	250	Guadalupian	249

A large number of volcanoes erupting over a long interval could lower global temperatures by injecting massive quantities of volcanic ash and dust into the atmosphere (Fig. 124). Heavy clouds of volcanic dust have a high albedo and reflect solar radiation back into space, thereby shading the Earth. The lowered global temperatures could cause mass extinctions of plants and animals by reducing the rate of global photosynthesis. However, plants can survive short-term catastrophes far better than animals by wilting during droughts, resprouting from roots, and prolonged inactivity as seeds.

Acid rain from extensive volcanic activity a hundred times more intense than at present could cause widespread destruction of terrestrial and marine species by defoliating plants and raising the acidity level in the ocean. Acid gases spewed into the atmosphere might deplete the ozone layer, allowing deadly ultraviolet radiation from the sun to bathe the planet. Volcanic eruptions also affect the climate by altering the composition of the atmosphere. Large volcanic eruptions spew so much ash and aerosols into the atmosphere they not only block sunlight but also absorb solar radiation, which heats the atmosphere, causing thermal imbalances and unstable climatic conditions.

Geologic evidence taken from sequences of volcanic rock on the ocean floor, which record the polarity of the Earth's magnetic field when they cool

Figure 124   A large eruption cloud from the July 22, 1980, eruption of Mount St. Helens. Courtesy of USGS

and solidify, shows that the geomagnetic field has reversed often in the past. After a long stable period of hundreds of thousands of years, the magnetic field strength gradually decays over a short period of several thousand years. At some point, it collapses altogether, and a short time later it regenerates with opposite polarity.

A comparison of geomagnetic field reversals with variations in the climate often shows a striking agreement (Table 8). Furthermore, certain magnetic reversals coincide with the extinction of species. Magnetic field reversals also might have been responsible for periods of glaciation. Reversals in the geomagnetic field and excursions of the magnetic poles appear to correlate with periods of rapid cooling and extinction of species. For example, the Gothenburg geomagnetic excursion occurred about 13,500 years ago in the midst of a longer period of rapid global warming toward the end of the last ice age. It resulted in plummeting temperatures and advancing glaciers for a thousand years, apparently caused by a weakened magnetic field.

TABLE 8   COMPARISON OF MAGNETIC REVERSALS WITH OTHER PHENOMENA
(dates in millions of years ago)

Magnetic Reversal	Unusual Cold	Meteorite Activity	Sea Level Drops	Mass Extinctions
0.7	0.7	0.7		
1.9	1.9	1.9		
2.0	2.0			
10				11
40			37–20	37
70			70–60	65
130			132–125	137
160			165–140	173

Ten or more major meteorites have impacted over the last 600 million years, some of which coincide with mass extinctions. When a large asteroid or comet slams into the Earth, a huge explosion lofts massive amounts of sediment into the atmosphere that shuts out the sun. Darkness might last several months, halting photosynthesis and eliminating near-surface phytoplankton in the ocean. The effects of these extinctions would cascade up the food chain, killing off both large and small marine and terrestrial species. A massive bombardment of meteoroids or comets might strip away the upper atmospheric ozone layer, leaving all species on the surface vulnerable to the sun's deadly ultraviolet rays. The exposure would kill terrestrial plants and animals along with primary producers in the surface waters of the ocean.

A controversial theory to explain the extinction of the dinosaurs contends that a 6- to 9-mile-wide asteroid struck the Earth and gouged out a crater 100 miles in diameter. Boundary rocks between the Cretaceous and Tertiary periods throughout the world contain a thin layer of fallout material composed of mud with shock-impact sediments, microspherules produced by impact melt, organic carbon from massive forest fires, the mineral stishovite found only at known impact sites, meteoritic amino acids, and an unusually high iridium content. The geologic record holds clues to other large meteorite impacts associated with iridium anomalies that coincide with extinction episodes. Therefore, giant impacts might have had a hand in the initiation and extermination of species throughout geologic history.

EFFECTS OF EXTINCTIONS

Since life first appeared on Earth, there have always been gradual die-outs of species, called background or normal extinctions. Major extinction events are punctuated by periods of lower extinction rates, and species have regularly come and gone even during optimum conditions. However, mass extinctions are not simply the intensification of processes operating during these background periods. Survival traits developed during times of lower extinction rates become irrelevant during mass extinctions. This suggests that mass extinctions might be less discriminatory with respect to the environment than normal extinctions. Different processes might be operating during times of mass extinction than those operating during normal extinctions. Moreover, the same types of species that succumb to mass extinctions also succumb to background extinctions—only a lot more of them.

Species that survive mass extinction are particularly hardy and resistant to subsequent random changes in the environment. They tend to occupy large geographic ranges that contain many groups of related species. But just because a species survives extinction does not always mean it was superior or better suited to its environment. Species that became extinct might have been developing certain unfavorable traits during background times. This could occur even within generations of the same organism, as daughter species develop better survival skills and replace their parent species. In this case, those characteristics that permit a species to live successfully during normal periods might become irrelevant when major extinction events occur.

After mass extinction, species that survive radiate outward to fill vacated habitats, spawning the development of entirely new species. These new species might develop novel adaptations that give them a survival advantage over other species. These adaptations might lead to exotic organisms that prosper during intervals of normal background extinctions, but because of overspecialization are incapable of surviving a mass extinction.

The geologic record seems to imply that nature is constantly experimenting with new forms of life. Once a species becomes extinct, it is lost forever, and the chances of its unique combination of genes reappearing are astronomical. Evolution seems to run in one direction, and although it perfects species to live at their optimum in their respective environments, it can never run backwards. However, convergent evolution does make it possible for a species to physically resemble an entirely different species if only they share similar environmental circumstances.

Extinction reduces the number of different species as well as the total number of species. Afterward, the biological system seems to be temporarily immune to random cataclysms. Species that survive mass extinction are particularly hardy and resilient toward subsequent environmental changes. Furthermore, after a major extinction event, few species are left to die out. Therefore, until many species have evolved, including extinction-prone types, any intervening catastrophes would have comparatively little effect.

After each extinction, the biological world requires a recovery period before it is again ready to face another major extinction event. Each time a mass extinction occurs, it resets the evolutionary clock, as though life were forced to start anew. After the great extinction that ended the Paleozoic, 240 million years ago, which left the world almost as devoid of species as when the era began, the Earth witnessed many remarkable advancements. Species that survived the extinction were similar to populations living today. Many of these same species survived the end-Cretaceous extinction, suggesting they might have perfected survival characteristics that other species lacked.

As the world was recovering from the extinction at the end of the Paleozoic, many regions of the ocean became filled with numerous specialized organisms, and the overall diversity of species rose to unprecedented heights. However, instead of evolving entirely novel forms, like those that evolved during the Cambrian explosion, species that survived the end-Paleozoic extinction developed morphologies based on simple skeletal types, with few experimental life forms.

MODERN EXTINCTIONS

Man came into existence during the greatest biological diversity in the history of the planet, when 70 percent of today's species evolved. We are a relatively new species on the geologic time scale, especially when compared to other species, some of which have been in existence for hundreds of millions of years. Within the past few thousand years, people have radiated into all lands, and within the last few hundred years, human populations have swelled a thousandfold. We are the most adaptable species, capable of living in diverse environments, often nudging out other species in competition.

Humans are the only creatures on Earth that have forced the extinctions of large numbers of other species. We have been called the "human volcano" because our influences on the environment are global just like major volcanic eruptions. Such convulsions have been cited as causes of mass extinction of species, and major changes in the Earth's critical cycles brought on by human interference could spell catastrophe for all mankind as well as the rest of the living world.

More than 90 percent of all species exist on land. For the first time in geologic history, plants are being extinguished in tragic numbers. If current trends continue, 7 percent of all plant species are likely to become extinct by the end of the century. In the United States alone, as much as 10 percent of the nation's plant species are destined for extinction. Plants are at risk

Figure 125   Exxon workers using high-pressure cold water to rinse crude oil from a beach in Prince William Sound, Alaska. Photo by Jill Bauermeister, courtesy of USGS

of extinction from forest destruction, expansion of agriculture, and the spread of urbanization. Over a thousand domestic species of plants and animals are either endangered or threatened with extinction. Plants and animals not directly beneficial to man are crowded out, as growing human populations continue to squander the Earth's space and resources and contaminate the environment with pollution (Fig. 125).

Everywhere humans have gone, they have wiped out entire species of mammals, birds, reptiles, fish, and other life forms, including indigenous peoples. Between 1600 and 1900, during a period of extensive maritime exploration, humans eliminated 75 known species, mostly birds and mammals. Passenger pigeons that once darkened the skies over North America by the billions became totally extinct by 1914. We continue to destroy life on every continent and island we inhabit, except on a much larger scale than our forebears, simply because there are so many more of us.

Fish species are rapidly disappearing throughout the world due to deforestation, which causes increased sedimentation, and acid rain, which acidifies lakes and streams. Nor are the oceans immune, as large fisheries are seriously depleted to feed growing human populations. Even the great sharks, which have been extremely successful predators for the last 400 million years, are succumbing to human overfishing.

Birds are also at risk, and humans have caused more than 1 percent of all bird species to become extinct over the last few centuries. Half the Hawaiian bird population has collapsed from overhunting and destruction of native forests since original Polynesian habitation some 1,600 years ago. In more recent times, the bird population has been further reduced by 15 percent. This situation is typical of the impact of human settlement on island communities, which are especially vulnerable.

An estimated 2,000 bird species, or about one-fifth of all bird species that existed a few thousand years ago, have fallen victim to prehistoric exterminations. Today, up to 20 percent of the bird species are again endangered or at imminent risk of extinction. Large, flightless birds, like the extinct dodo and great auk, are particularly at risk from human interference. Island birds are frequently flightless because they no

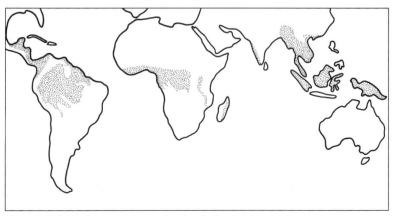

Figure 126   Location of tropical rain forests.

longer need to take to the air to escape predators. Species inhabiting islands are also extremely vulnerable to humans simply because they have nowhere to flee. Island animals often develop unique characteristics apart from their mainland relatives that make them particularly vulnerable to humans.

Tropical forests in the New World have shrunk by about one-third, and those of Africa have been reduced by as much 75 percent since 1960 (Fig. 126). These regions are the wintering grounds for migratory birds from the Northern Hemisphere. The disappearance of the rain forests could mean a decline in bird populations for the northern countries besides countless other species that inhabit the forests themselves. Already, birds are starting to disappear in alarming numbers. Like canaries used to detect poisonous gas in coal mines, they could be offering a warning that our planet is in trouble.

THE WORLD AFTER

As a possible prelude to global extinction, frogs and other amphibians that have been in existence for over 300 million years are disappearing at alarming rates all over the world. Moreover, amphibians are vanishing from nature preserves, where little human perturbation occurs. These creatures might be sounding an early warning that the Earth is grave danger.

Species are also becoming extinct from overhunting, introduced species that prey on or directly compete with indigenous species, destruction of habitats by deforestation (Fig. 127) along with other destructive human activi-

Figure 127 **A smoke cloud from forest fires that obscured almost a third of South America in the fall of 1989.** Courtesy of NASA

Figure 128   A view of Earth over the Andes Mountains from space. Courtesy of NASA

ties, and the collapse of food chains. Every species depends on other species for its survival, and when too many species become extinct in an ecosystem the remaining species are at risk of extinction by the "domino effect." If this process were to continue globally due to destructive human activities, it could initiate the collapse of the biosphere and incur one of the worst extinction events in the history of our planet.

When viewed from space the Earth appears to be a living entity filled with luxuriant growth (Fig. 128). But in reality, it is losing precious species of plants and animals at an alarming rate, due mainly to human destruction of their habitats. Throughout the world, the die-out of species is thousands of times greater than the natural background extinction rate prior to the appearance of man. Over 90 percent of all species have yet to be described and therefore are unknown to science. Biologists are in a desperate race to classify as many species as possible before many more are gone. Simple creatures, like bacteria, which make it possible for all other species to survive, comprise over 80 percent of the Earth's biomass (the total weight of living matter). They are absolutely essential to the health of the planet and if they disappear, so do we.

If current trends continue, by the middle of the next century the number of extinct species could exceed those lost in the great extinctions of the geologic past. By our very actions, we are upsetting the delicate balance of nature, and should it tilt ever so slightly, cataclysmic changes could result. Mass extinctions normally occur over periods of thousands or even millions of years, but because of our disturbance the extinction of large numbers of species would take place in a mere century. We have yet to feel the adverse effects of such a huge die-out of species. However, once a wave of extinction is set in motion, it ultimately undermines the quality or even the possibility of human life.

GLOSSARY

aa lava   a lava that forms large, jagged, irregular blocks

abrasion   erosion by friction, generally caused by rock particles carried in running water, ice, or wind

agglomerate   a pyroclastic rock composed of consolidated volcanic fragments

albedo   the amount of sunlight reflected from an object, determined by its color

alluvium   stream-deposited sediment

alpine glacier   a mountain glacier or a glacier in a mountain valley

andesite   an intermediate type of volcanic rock between basalt and rhyolite

anticline   folded sediments that slope downward and away from a central axis

Apollo asteroid   an asteroid from the main belt between Mars and Jupiter that crosses the Earth's orbit

aquifer   a subterranean bed of sediments through which groundwater flows

ash fall   the fallout of small, solid particles from a volcanic eruption cloud

asteroid	a rocky or metallic body orbiting the sun between Mars and Jupiter; possibly once part of a larger body that disintegrated
asteroid belt	a band of asteroids orbiting the sun between the orbits of Mars and Jupiter
asthenosphere	a layer of the upper mantle, roughly between 60 and 200 miles below the surface, that is more plastic than the rock both above and below and might be in convective motion
astrobleme	eroded remains of an ancient impact structure produced by a large cosmic body
back-arc basin	a seafloor spreading system of volcanoes caused by extension behind an island arc that is above a subduction zone
basalt	a dark volcanic rock that is usually quite fluid in the molten state
batholith	the largest of the intrusive igneous bodies, at more than 40 square miles on its uppermost surface
bedrock	solid layers of rock lying beneath younger material
black smoker	tube created as superheated hydrothermal water rising to the surface at a midocean ridge. The water is supersaturated with metals, and when exiting through the seafloor quickly cools as the dissolved metals precipitate, resulting in black, smokelike effluent.
blowout	a hollow caused by wind erosion
bolide	an exploding meteor whose fireball is often accompanied by a bright light and sound when passing through the atmosphere
caldera	a large pitlike depression at the summits of some volcanoes formed by great explosive activity and collapse
calving	formation of icebergs by ice breaking off of glaciers that enter the ocean
carbonaceous chondrites	stony meteorites that contain abundant organic compounds

chondrite	the most common type of meteorite, composed mostly of rocky material with small spherical grains called chondrules
circum-Pacific belt	active seismic regions around the rim of the Pacific plate that coincide with the Ring of Fire
cirque	a glacial erosional feature, producing an amphitheaterlike head of a glacial valley
comet	a celestial body believed to originate from a cloud that surrounds the sun. A comet develops a long tail of gas and dust particles when traveling near the inner Solar System.
conduit	a passageway leading from a reservoir of magma to the surface of the Earth through which volcanic products pass
continent	a landmass composed of light, granitic rock that rides on denser rocks of the upper mantle
continental drift	the concept that the continents have been drifting across the surface of the Earth throughout geologic time
continental glacier	an ice sheet covering a portion of a continent
continental shelf	the offshore area of a continent in shallow sea
continental shield	ancient crustal rocks upon which the continents grew
continental slope	the transition from the continental margin to the deep-sea basin
convection	a circular, vertical flow of fluid medium due to heating from below. As materials are heated, they become less dense and rise, while cooler, heavier materials sink.
cordillera	a range of mountains that includes the Rockies, Cascades, and Sierra Nevada in North America and the Andes in South America
core	the central part of a planet and consisting of a heavy iron-nickel alloy
craton	the stable interior of a continent, usually composed of the oldest rocks in the continent

crevasse	a deep fissure in the Earth or in a glacier
crust	the outer layers of a planet or a moon's rocks
crustal plate	one of several lithospheric plates comprising the Earth's surface rocks
delta	a wedge-shaped layer of sediments deposited at the mouth of a river
diapir	the buoyant rise of a molten rock through heavier rock
divergent plate boundary	the boundary between crustal plates where the plates move apart. Generally corresponds to the midocean ridges where new crust is formed by the solidification of liquid rock rising below.
drought	a period of abnormally dry weather so extensive that the lack of water causes deleterious effects on agriculture and other biological activities
drumlin	an elongated hill of glacial debris pointing in the direction of glacial movement
dune	a ridge of windblown sediments usually in motion
earthquake	the sudden breaking of rocks in response to geologic forces within the Earth
East Pacific Rise	midocean spreading system that runs north-south along the eastern side of the Pacific. The predominant location upon which the hot springs and black smokers have been discovered.
elastic rebound theory	the theory that earthquakes depend on rock elasticity
eolian	a deposit of windblown sediment
epicenter	the point on the Earth's surface directly above the focus of an earthquake
erratic	a glacially deposited boulder far from its source
escarpment	a mountain wall caused by elevation of a block of land
esker	a curved ridge of glacially deposited material

extrusive	an igneous volcanic rock ejected onto the surface of a planet or moon
fault	a breaking of crustal rocks caused by Earth movements
fissure	a large crack in the crust through which magma might escape from a volcano
fluvial	pertaining to river deposits
focus	the point of origin of an earthquake; also called a hypocenter
formation	a combination of rock units that can be traced over distance
frost heaving	the lifting of rocks to the surface by the expansion of freezing water
frost polygons	polygonal patterns of rocks formed by repeated freezing
fumarole	a vent through which steam or other hot gases escape from underground, such as a geyser
geothermal	the generation of hot water or steam by hot rocks in the Earth's interior
geyser	a spring that ejects intermittent jets of steam and hot water
glacier	a thick mass of moving ice occurring where winter snowfall exceeds summer melting
Gondwana	a southern supercontinent of Paleozoic time, consisting of Africa, South America, India, Australia, and Antarctica. It broke up into the present continents during the Mesozoic era.
graben	a valley formed by a downdropped fault block
granite	a coarse-grained, silica-rich rock, consisting primarily of quartz and feldspars. It is the principal constituent of the continents and believed to be derived from a molten state beneath the Earth's surface.
gravity fault	motion along a fault plane that moves as if pulled downslope by gravity; also called a normal fault

groundwater	the water derived from the atmosphere that percolates and circulates below the surface of the Earth
guyot	an undersea volcano that reached the surface of the ocean, whereupon its top was flattened by erosion. Later, subsidence caused the volcano to sink below the surface preserving its flat-top appearance.
haboob	a violent dust storm or sandstorm
horn	a peak on a mountain formed by glacial erosion
horst	an elongated, uplifted block of crust bounded by faults
hot spot	a volcanic center that has no relation to a plate boundary location; an anomalous magma generation site in the mantle
hydrothermal	relating to hot water in the crust, often in convective motion
hypocenter	the point of origin of earthquakes; also called focus
Iapetus Sea	a former sea that occupied an area similar to the present Atlantic Ocean prior to the assemblage of Pangaea
ice age	a period of time when large areas of the Earth are covered by glaciers
iceberg	a portion of a glacier broken off, or calved, upon entering the sea
ice cap	a polar cover of ice and snow
igneous rocks	all rocks that have solidified from a molten state
ignimbrite	volcanic deposit created by ejections of incandescent solid particles
impact	the point on the surface upon which a celestial object lands
interglacial	a warming period between glaciations
intrusive	any igneous body that has solidified in place below the Earth's surface
island arc	volcanoes landward of a subduction zone parallel to a trench and above the melting zone of a subducting plate

GLOSSARY

karst	a terrain comprised of numerous sinkholes in limestone
kettle	a depression in the ground caused by a buried block of glacial ice
lahar	a hot mudflow or ashflow on the slopes of a volcano
landslide	rapid downhill movement of earth materials often triggered by earthquakes
lapilli	small, solid pyroclastic fragments
Laurasia	the northern supercontinent of Paleozoic time consisting of North America, Europe, and Asia
lava	molten magma after it has flowed out onto the surface
liquefaction	the liquefying of subterranean sediment layers due to earthquake activity
lithosphere	a rigid outer layer of the mantle, typically about 60 miles thick, overridden by the continental and oceanic crusts and divided into segments called plates
loess	a thick deposit of airborne dust
magma	a molten rock material generated within the Earth; the constituent of igneous rocks, including volcanic eruptions
magnitude scale	a scale for rating earthquake energy
mantle	the part of a planet below the crust and above the core, composed of dense rocks that might be in convective flow
mass wasting	the downslope movement of rock under the direct influence of gravity
megaplume	a large volume of mineral-rich warm water above an oceanic rift
meteor	a small celestial body that becomes visible as a streak of light when entering the Earth's atmosphere
meteorite	a metallic or stony body from space that enters the Earth's atmosphere and impacts on the surface

meteoritic crater	a depression in the crust produced by the bombardment of a large meteorite
meteoritics	the science that deals with meteors and related phenomena
meteoroid	a meteor in orbit around the sun
meteor shower	a phenomenon observed when large numbers of meteors enter the Earth's atmosphere. Their luminous paths appear to diverge from a single point.
microearthquake	a small earth tremor
micrometeorites	small, grain-sized bodies
microtektites	small, spherical grains created by the melting of surface rocks during a large meteorite impact
Mid-Atlantic Ridge	the seafloor spreading ridge of volcanoes that marks the extensional edge of the North American and South American plates to the west and the Eurasian and African plates to the east
midocean ridge	a submarine ridge along a divergent plate boundary where a new ocean floor is created by the upwelling of mantle material
moraine	a ridge of erosional debris deposited by the melting margin of a glacier
mountain roots	the deeper crustal layers under mountains
normal fault	a gravity fault in which one block of crust slides down another block of crust along a steeply tilted plane
nu'ée ardente	a volcanic pyroclastic eruption of hot ash and gas
orogeny	a process of mountain building by tectonic activity
outgassing	the loss of gas from within a planet as opposed to degassing, the loss of gas from meteorites
overthrust	a thrust fault in which one segment of crust overrides another segment of crust for a great distance
pahoehoe lava	a lava that forms ropelike structures when cooled

Pangaea	ancient supercontinent that included all the lands of the Earth
Panthalassa	the great world ocean that surrounded Pangaea
peridotite	the most common ultramafic rock type in the Earth's mantle from which basalt is formed
permafrost	permanently frozen ground in the arctic regions
permeability	the ability to transfer fluid through cracks, pores, and interconnected spaces within a rock
pillow lava	lava extruded on the ocean floor giving rise to tabular shapes
placer	a deposit of rocks left behind by a melting glacier; any ore deposit that is enriched by stream action
plate tectonics	the theory that accounts for the major features of the Earth's surface in terms of the interaction of lithospheric plates
pluton	an underground body of igneous rock younger than the rocks that surround it, formed where molten rock oozes into a space between older rocks
porosity	the percentage of pore spaces in a rock between crystals and grains, usually filled with water
pumice	volcanic ejecta that is full of gas cavities and extremely lightweight
pyroclastics	the fragmental ejecta released explosively from a volcanic vent
reef	biologic community that lives at the edge of an island or continent; the shells from organisms form a limestone deposit
regression	a fall in sea level, exposing continental shelves to erosion
resurgent caldera	a large caldera that experiences renewed volcanic activity, doming up the caldera floor
ridge crest	an axis of midocean volcanoes aligned along the edge of two plates extending away from each other

rift valley	the center of an extensional spreading center where continental or oceanic plate separation occurs
rille	a trench formed by a collapsed lava tunnel
saltation	the movement of sand grains by wind or water
scarp	a steep slope formed by Earth movements
seafloor spreading	the theory that the ocean floor is created by the separation of lithospheric plates along the midocean ridges, with new oceanic crust formed from mantle material that rises from the mantle to fill the rift
seamount	a submarine volcano
shield	the exposed Precambrian nucleus of a continent
shield volcano	a broad, low-lying volcanic cone built up by lava flows of low viscosity
sinkhole	a large pit formed by the collapse of surface materials undercut by the dissolution of subterranean limestone
spherules	small, spherical glassy grains found on certain types of meteorites, lunar soils, and at large meteorite impact sites
stishovite	a quartz mineral produced by extremely high pressures, such as those generated by a large impact
stratovolcano	an intermediate volcano characterized by a stratified structure from alternating emissions of lava and fragments
strewn field	usually large area where tektites from a large meteorite impact are found
striation	scratches on bedrock made by rocks embedded in a moving glacier
subduction zone	an area where an oceanic plate dives below a continental plate into the mantle. Ocean trenches are the surface expression of a subduction zone.
submarine canyon	a deep gorge residing undersea and formed by the underwater extension of rivers

subsidence	the collapse of sediments due to the removal of underground fluids
surge glacier	a continental glacier that heads toward the sea at a high rate of advance during certain periods
syncline	a fold in which the beds slope inward toward a common axis
tectonic activity	the formation of the Earth's crust by large-scale Earth movements throughout geologic time
tektites	small, glassy minerals created by the melting of surface rocks by an impact of a large meteorite
tephra	all clastic material from dust particles to large chunks, expelled from volcanoes during eruptions
Tethys Sea	the hypothetical mid-latitude area of the oceans separating the northern and southern continents of Laurasia and Gondwana several hundred million years ago
till	sedimentary material deposited by glacial ice
tillite	a sedimentary deposit composed of glacial till
transform fault	a fracture in the Earth's crust along which lateral movement occurs and a common feature of midocean ridges
transgression	a rise in sea level that causes flooding of the shallow edges of continental margins
trench	a depression on the ocean floor caused by plate subduction
tsunami	a sea wave generated by an undersea earthquake or marine volcanic eruption
tuff	a volcanic rock formed by pyroclastic fragments
tundra	permanently frozen ground at high latitudes
turbidite	a slurry of mud that periodically slides down often gentle submarine slopes toward the deep seafloor
volcanic ash	fine pyroclastic material injected into the atmosphere by an erupting volcano
volcanic bomb	a solidified blob of molten rock ejected from a volcano

volcanic cone	the general term applied to any volcanic mountain with a conical shape
volcanic crater	the inverted conical depression found at the summit of most volcanoes and formed by the explosive emission of volcanic ejecta
volcanism	any type of volcanic activity, including volcanoes, geysers, and fumaroles
volcano	a fissure or vent in the crust through which molten rock rises to the surface to form a mountain

BIBLIOGRAPHY

THE DYNAMIC EARTH

Anderson, Don L. "The Earth as a Planet: Paradigms and Paradoxes." *Science* 223 (January 27, 1984): 347–354.

Bercovici, Dave, Gerald Schubert, and Gary A. Glatzmaier. "Three-Dimensional Spherical Models of Convection in the Earth's Mantle." *Science* 244 (May 26, 1989): 950–954.

Bonatti, Enrico. "The Rifting of Continents." *Scientific American* 256 (March 1987): 97–103.

Courtillot, Vincent and Gregory E. Vink. "How Continents Break Up." *Scientific American* 249 (July 1983): 43–49.

Francheteau, Jean. "The Oceanic Crust." *Scientific American* 249 (September 1983): 114–129.

Gordon, Richard G. and Seth Stein. "Global Tectonics and Space Geodesy." *Science* 256 (April 17, 1992): 333–341.

Macdonald, Kenneth C. and Paul J. Fox. "The Mid-Ocean Ridge." *Scientific American* 262 (June 1990): 72–79.

Monastersky, Richard. "The Whole-Earth Syndrome." *Science News* 133 (June 11, 1988): 378–380.

Murphy, J. Brendan and R. Damian Nance. "Mountain Belts and the Supercontinent Cycle." *Scientific American* 266 (April 1992): 84–91.

Peacock, Simon M. "Fluid Processes in Subduction Zones." *Science* 248 (April 20, 1990): 329–336.

Wickelgren, Ingrid. "Simmering Planet." *Discover* 11 (July 1990): 73–75.

EARTHQUAKES

Bolt, Bruce A. "Balance of Risks and Benefits in Preparation for Earthquakes." *Science* 251 (January 11, 1991): 169–174.

Finkbeiner, Ann. "California's Revenge." *Discover* 11 (September 1990): 79–85.

Fischman, Joshua. "Falling into the Gap." *Discover* 13 (October 1992): 57–63.

Frohlich, Cliff. "Deep Earthquakes." *Scientific American* 260 (January 1989): 48–55.

Gore, Rick. "Our Restless Planet." *National Geographic* 168 (August 1985): 142–179.

Heaton, Thomas H. and Stephen H. Hartzell. "Earthquake Hazards on the Cascadia Subduction Zone." *Science* 236 (April 10, 1987): 162–168.

Jayaraman, K. S. "India Seeks to Learn the Lessons of the Maharashta Earthquake." *Nature* 365 (October 14, 1993): 593.

Johnston, Arch C. "A Major Earthquake Zone on the Mississippi." *Scientific American* 246 (April 1982): 60–68.

Johnston, Arch C. and Lisa R. Kanter. "Earthquakes in Stable Continental Crust." *Scientific American* 262 (March 1990): 68–75.

Roman, Mark B. "Finding Fault." *Discover* 9 (August 1988): 57–63.

Stein, Ross S. and Robert S. Yeats. "Hidden Earthquakes." *Scientific American* 260 (June 1989): 48–57.

Unklesbay, A. G. "Midwest Earthquakes." *Earth Science* 40 (Winter 1987): 11–13.

Vogel, Shawna. "Shocks Heard Round the World." *Discover* 11 (January 1990): 68–70.

Wesson, Robert L. and Robert E. Wallace. "Predicting the Next Great Earthquake in California." *Scientific American* 252 (February 1985): 35–43.

VOLCANIC ERUPTIONS

Berreby, David. "Barry versus the Volcano." *Discover* 12 (June 1991): 60–67.

Chen, Allan. "The Thera Theory." *Discover* 10 (February 1989): 77–83.

Decker, Robert and Barbara Decker. "The Eruptions of Mount St. Helens." *Scientific American* 244 (March 1981): 68–80.

Dvorak, John J., Carl Johnson, and Robert I. Tilling. "Dynamics of Kilauea Volcano." *Scientific American* 267 (August 1992): 46–53.

Kerr, Richard A. "Volcanoes: Old, New, and—Perhaps—Yet to Be." *Science* 250 (December 21, 1990): 1660–1661.

LaMarche, Valmore C., Jr., and Katherine K. Hirschboeck. "Frost Rings in Trees as Records of Major Volcanic Eruptions." *Nature* 307 (January 12, 1984): 121–126.

Rampino, Michael R. and Richard B. Strothers. "Flood Basalt Volcanism during the Past 250 Million Years." *Science* 241 (August 5, 1988): 663–667.

Roche, Kirby. "The Mystique of Disaster." *Weatherwise* 43 (October 1990): 262–264.

Stommel, Henry and Elizabeth Stommel. "The Year without a Summer." *Scientific American* 240 (June 1979): 176–186.

Stothers, Richard B. "The Great Tambora Eruption in 1815 and Its Aftermath." *Science* 224 (June 15, 1984): 1191–1197.

EARTH MOVEMENTS

"Facing Geologic and Hydrologic Hazards." *U.S. Geological Survey Professional Paper 1240-B*. Washington DC: Government Printing Office, 1981.

Friedman, Gerald M. "Slides and Slumps." *Earth Science* 41 (Fall 1988): 21–23.

Gibbons, Boyd. "Do We Treat Our Soil Like Dirt?" *National Geographic* 166 (September 1984): 353–388.

Monastersky, Richard. "When Mountains Fall." *Science News* 142 (August 29, 1992): 136–138.

———. "Spotting Erosion from Space." *Science News* 136 (July 22, 1989): 61.

———. "Soil May Signal Imminent Landslide." *Science News* 134 (November 12, 1988): 318.

Norris, Robert M. "Sea Cliff Erosion." *Geotimes* 35 (November 1990): 16–17.

Peterson, Ivars. "Digging into Sand." *Science News* 136 (July 15, 1989): 40–42.

Shaefer, Stephen J. and Stanley N. Williams. "Landslide Hazards." *Geotimes* 36 (May 1991): 20–22.

Zimmer, Carl. "Landslide Victory." *Discover* 12 (February 1991): 66–69.

CATASTROPHIC COLLAPSE

Bolton, David W. "Underground Frontiers." *Earth Science* 40 (Summer 1987): 16–18.

Francis, Peter. "Giant Volcanic Calderas." *Scientific American* 248 (June 1983): 60–70.

Francis, Peter and Stephen Self. "Collapsing Volcanoes." *Scientific American* 256 (June 1987): 91–97.

Holzer, T. L., T. L. Youd, and T. C. Hanks. "Dynamics of Liquefaction during the 1987 Superstition Hills, California Earthquake." *Science* 244 (April 7, 1989): 56–59.

Kerr, Richard A. "Delving into Faults and Earthquake Behavior." *Science* 235 (January 9, 1987): 165–166.

Lipske, Mike. "Wonder Holes." *International Wildlife* 20 (February 1990): 47–51.

Marsden, Sullivan S., Jr., and Stanley N. Davis. "Geological Subsidence." *Scientific American* 216 (June 1967): 93–100.

Simon, Cheryl. "A Giant's Troubled Sleep." *Science News* 124 (July 16, 1983): 40–41.

Weisburd, Stefi. "Sensing the Voids Underground." *Science News* 130 (November 22, 1986): 329.

FLOODS

Adler, Jerry. "Troubled Waters." *Newsweek* (July 26, 1993): 21–26.

Ambroggi, Robert P. "Water." *Scientific American* 243 (September 1980): 101–115.

Cathles, Lawrence M., III. "Scales and Effects of Fluid Flow in the Upper Crust." *Science* 248 (April 20, 1990): 323–328.

Ingersoll, Andrew P. "The Atmosphere." *Scientific American* 249 (September 1983): 162–174.

Keller, Edward A. *Environmental Geology*. Columbus: Merrill, 1976.

Macilwain, Colin. "Conservationists Fear Defeat on Revised Flood Control Policies." *Nature* 365 (October 7, 1993): 478.

Ramage, Colin S. "El Niño." *Scientific American* 254 (June 1986): 77–83.

Rasmusson, Eugene M. "Meteorological Aspects of the El Niño/Southern Oscillation." *Science* 222 (December 16, 1983): 1195–1202.

Webster, Peter J. "Monsoons." *Scientific American* 245 (August 1981): 109–118.

DUST STORMS

Abelson, Philip H. "Climate and Water." *Science* 243 (January 27, 1989): 461.

Bower, Bruce. "Shuttle Radar Is Key to Sahara's Secrets." *Science News* 125 (April 21, 1984): 244.

Idso, Sherwood B. "Dust Storms." *Scientific American* 235 (October 1976): 108–114.

Manabe, S. and R. T. Wetherald. "Reduction in Summer Soil Wetness Induced by an Increase in Atmospheric Carbon Dioxide." *Science* 232 (May 2, 1986): 626–628.

Maranto, Gina. "A Once and Future Desert." *Discover* 6 (June 1985): 32–39.

Pennisi, Elizabeth. "Dancing Dust." *Science News* 142 (October 3, 1992): 218–220.

Raloff, Janet. "Salt of the Earth." *Science News* 126 (November 10, 1984): 298–301.

Repetto, Robert. "Deforestation in the Tropics." *Scientific American* 262 (April 1990): 36–42.

Schneider, Stephen H. "Climate Modeling." *Scientific American* 256 (May 1987): 72–80.

White, Robert M. "The Great Climate Debate." *Scientific American* 263 (July 1990): 36–43.

GLACIATION

Barnes-Svarney, Patricia. "Hubbard Glacier." *Earth Science* 40 (Fall 1987): 20.

Beard, Jonathan. "Glaciers on the Run." *Science 85* 6 (February 1985): 84–85.

Bowen, D. Q. "Antarctic Ice Surges and Theories of Glaciation." *Nature* 283 (February 14, 1980): 619–621.

Broecker, Wallace S. and George H. Denton. "What Drives Glacial Cycles?" *Scientific American* 262 (January 1990): 49–56.

Gordon, Arnold L. and Josefino C. Comiso. "Polynyas in the Southern Ocean." *Scientific American* 258 (June 1988): 90–97.

Hansen, J. E. "Global Sea Level Trends." *Nature* 313 (January 31, 1985): 349–350.

Mathews, Samuel W. "Ice on the World." *National Geographic* 171 (January 1987): 84–103.

Meyer, Alfred. "Between Venice and the Deep Blue Sea." *Science 86* 7 (July/August 1986): 50–57.

Mollenhauer, Erik and George Bartunek. "Glaciers on the Move." *Earth Science* 41 (Spring 1988): 21–24.

Parfit, Michael. "Antarctic Meltdown." *Discover* 10 (September 1989): 39–47.

Radok, Uwe. "The Antarctic Ice." *Scientific American* 253 (August 1985): 98–105.

ASTEROID COLLISIONS

Alvarez, Walter and Frank Asaro. "An Extraterrestrial Impact." *Scientific American* 263 (October 1990): 78–84.

Binzel, Richard P., M. Antonietta, and Marcello Fulchignoni. "The Origins of the Asteroids." *Scientific American* 265 (October 1991): 88–94.

Gehrels, Tom. "Asteroids and Comets." *Physics Today* 38 (February 1985): 33–41.

Grieve, Richard A. F. "Impact Cratering on the Earth." *Scientific American* 262 (April 1990): 66–73.

Hildebrand, Alan R. and William V. Boynton. "Cretaceous Ground Zero." *Natural History* (June 1991): 47–52.

Mathews, Robert. "A Rocky Watch for Earthbound Asteroids." *Science* 255 (March 6, 1992): 1204–1205.

McFadden, Lucy A., Michael J. Gaffey, and Thomas B. McCord. "Near-Earth Asteroids: Possible Sources from Reflectance Spectroscopy." *Science* 229 (July 12, 1985): 160–162.

Morrison, David. "Target Earth: It Will Happen." *Sky & Telescope* 79 (March 1990): 261–265.

O'Keefe, John A. "The Tektite Problem." *Scientific American* 239 (August 1978): 116–125.

Sharpton, Virgil L. "Glasses Sharpen Impact Views." *Geotimes* 33 (June 1988): 10–11.

Weissman, Paul R. "Are Periodic Bombardments Real?" *Sky & Telescope* 79 (March 1990): 266–270.

MASS EXTINCTIONS

Alvarez, Luis W. "Mass Extinctions Caused by Large Bolide Impacts." *Physics Today* 40 (July 1987): 24–33.

Courtillot, Vincent E. "A Volcanic Eruption." *Scientific American* 263 (October 1990): 85–92.

Crowley, Thomas J. and Gerald R. North. "Abrupt Climate Change and Extinction Events in Earth History." *Science* 240 (May 20, 1988): 996–1001.

BIBLIOGRAPHY

Diamond, Jared. "Playing Dice with Megadeath." *Discover* 11 (April 1990): 55–59.

Ehrlich, Paul R. and Edward O. Wilson. "Biodiversity Studies: Science and Policy." *Science* 253 (August 16, 1991): 758–761.

Hallam, Anthony. "End-Cretaceous Mass Extinction Event: Argument for Terrestrial Causation." *Science* 238 (November 27, 1987): 1237–1241.

Levington, Jeffrey S. "The Big Bang of Animal Evolution." *Scientific American* 267 (November 1992): 84–91.

May, Robert M. "How Many Species Inhabit the Earth?" *Scientific American* 267 (October 1992): 42–48.

Raup, David M. "Biological Extinction in Earth History." *Science* 231 (March 28, 1986): 1528–1533.

Russell, Dale A. "The Mass Extinctions of the Late Mesozoic." *Scientific American* 256 (January 1982): 58–65.

Stanley, Steven M. "Mass Extinctions in the Ocean." *Scientific American* 250 (June 1984): 64–72.

Weisburd, Stefi. "Volcanoes and Extinctions: Round Two." *Science News* 131 (April 18, 1987): 248–250.

INDEX

Boldface page numbers indicate extensive treatment of a topic. *Italic* page numbers indicate illustrations or captions. Page numbers followed by m indicate maps; t indicate tables; g indicate glossary.

A

aa lava 167g
abrasion 167g
accretionary wedge 13
acid rain 82, 145, 148, 157, 162
Adams, Mount *see* Mount Adams
adaptation 160
Aden, Gulf of 14
Adriatic Sea 76
Afghanistan 22
Africa *see also specific country* (*e.g.,* Algeria); *geographic feature* (*e.g.,* Sahara Desert)
  droughts 117
  earthquakes 22
  extinctions 156t, 163
  geological dynamics 3m, 4, 120
  soil erosion 65
  volcanic eruptions 46
African plate 3m, 120
aftershock 32
agglomerate 167g
agriculture 64–67, 116–118, 162 *see also specific crop* (*e.g.,* grain)
Alabama 82

Alaska
  earthquakes *20,* 21–22, *33,* 63, 69, *70, 72–73, 74, 77, 130,* 131
  extinctions *161*
  floods *86*
  glaciers 120, 123, *130,* 130–131
  ground failures 69, 72–73, 77
  landslides 56, 69, *70,* 73, *74*
  lava tunnels 83
  rockslides 59
  soil slides *60*
  submarine slides 63
  subsidence 21, 69, 77
Alaska, Gulf of 21, 131
albedo 106t, 167g
  deserts 110–111
  extinctions 157
  glaciers 120, 124, *126,* 126–127, 131
Alberta, Canada rockslide (1903) 59
Aleutian Trench 22, 44
algae 152 *see also* diatoms
Algeria 28
alluvium 167g

Alpine glacier 123, 167g
alpine orogeny 22
Alps (Europe) 124, *126*
Amazon River and Basin (South America) 98, *111,* 112
Amirante Basin (Indian Ocean) 145
ammonites 153–154, *154*
amphibians 163
Anchorage, Alaska earthquake (1964) *see* Good Friday, Alaska earthquake (1964)
andesite 11, 44, 167g
Andes Mountains (South America) 3, 11, 22, 44, 57, 124, *164*
angle of repose 53
Antarctica
  craters 143
  dust storms 106
  glaciers 119–122, *121,* 124–125t, *128,* 129m, 129–132
  volcanic eruptions 156t
Antarctic plate 3m, 120
anticline 28, 167g

Apollo asteroids 141, 167g
Appalachian Mountains 53
aquifer 70, 167g
Arabia *see* Saudi Arabia
Arabian Desert 105m, 113
Arabian plate 3m
archaic mammals 135, 154
Arctic Ocean and Basin 120–121, 123, 125t
Arizona
  calderas 79
  craters 135
  dust storms 113
  flood control *101*
  floods 85t
  subsidence 76
Arkansas
  floods 84t–85t
  meteorites 144
Armenian earthquake (1988) 18, 26
Armero, Colombia 41, 63
ash fall 167g
Asia *see also specific country* (*e.g.,* Japan); *geographic feature* (*e.g.,* Himalaya Mountains)
  deserts 105m, 113
  glaciers 124
  meteorites 146
Assam, India earthquake (1897) 18
asteroid belt 141, 168g
asteroids x, 133, **140–148**, *141*, 142m, 158–159, 168g
asthenosphere 1, 6m, 7, 10, 168g
astroblemes 134, 168g
Atlantic Ocean and Basin 4, 9, 43, 148m
Augustine, Mount *see* Mount Augustine
auks (extinct birds) 162
Australia
  craters 136

deserts 105m
dust bowls 116
dust storms 113
earthquakes 22
geological dynamics 3m
glaciers 120, 124
volcanoes 44
Australian plate 3m, 44
Austria 124
avalanches 57, *130*
Azores 34

B

back-arc basin 11, 168g
Bahama Islands 82
Baker, Mount *see* Mount Baker
Bangladesh 89, 112
basalt 168g
  extinctions 156t
  geological dynamics 2, 4
  plate interaction 14
  resurgent calderas 80
  seafloor spreading 8
  subduction zones 10
  volcanoes 42–45m
Basin and Range Province (Weastern U.S.) 27
batholiths 43, 168g
Bay of Bengal *see* Bengal, Bay of
beaches *see* seashore
bearing strength loss 71–72
bedrock 168g
Bengal, Bay of 89, 112
Big Bear, California earthquake (1992) 24
Big Thompson River, Colorado flood (1976) 85t, *88*, 88–89
birds 162–163
Black Hills, South Dakota floods (1972) 85t
black smokers 9, 168g
blowouts 49, 114, 168g
blue holes 82
bolides 145, 168g
booming sands 115

Boston, Massachusetts 30
brachiopods 152–153
Brahmaputra River (India) 98, 112
bristlecone pine 103
Broken Bow, Nebraska 136
brontosaurs *153*

C

calcite 81
Calcutta, India earthquake (1737) 18
calderas 36–38, *41*, 49, **79–81**, 168g
California
  craters 50–51
  deserts 25, 103
  earthquakes *19*, 20, 22–28, *24, 26, 31, 69, 72, 73,* 77
  extinctions *151*
  floods 85t, *96*
  geological dynamics 3, *4*, 13–15
  ground failures *69, 71, 72,* 77
  irrigation *108*
  landslides 54, *54*
  lava tunnels 83
  mudflows *61*, 62
  resurgent calderas *80m*
  subsidence 75–77
  volcanic eruptions 40, 43–44, 49–51, *50*
calving 168g
Cambrian period 150t, 151, 160
Cameroon 48
Canada
  craters 134–136m
  deserts 118
  earthquakes 22
  floods 85t, 89, 92
  glaciers 123, 131
  rockslides 59
  submarine slide 63
Cape Canaveral National Seashore (Florida) *132*

carbonaceous chondrites 168g

carbon dioxide 46m, 48, 128

Carboniferous period 150t

carbon monoxide 48

Caribbean plate 3m

Cascade Range (California/Oregon/Washington) 3, *4*, 13, 40, 44, 49–51, 62

Cascadia subduction zone 13, 22, 44

Catania, Sicily 37

cattle grazing 111

Caucasus Mountains (Commonwealth of Independent States) 22

caverns 81–83

Cenozoic era 120, 154

Central America *see specific country (e.g., Mexico)*

Central U.S. floods (1957) 85t

cephalopods 154

Chad 103

Chad, Lake *see* Lake Chad

Charleston, South Carolina earthquake (1886) *29*, 30

Chicago, Illinois 30

Chicxulub structure 135

Chilean earthquake (1960) 13, 21, 28, 34

Chilean Trench 11, 28, 44

China
dust storms 113
earthquakes 16–17t, *17*, 18, 22, 60, 72
floods 95

chondrite 169g

cinder cone 14, 47

circum-Antarctic current 122

circum-Pacific belt 2, 21, 169g

cirque 169g

Clear Lake Volcanoes (California) 50

Cleveland, Ohio 75

climate *see* weather patterns

Climatic Optimum 103

Coalinga, California earthquake (1983) 25

coal mining 59, 83

Coast Range 25

Cocos plate 3m

Colombia 41, 63

Colorado
dust storms 113
earthquakes 75
floods 85t, *88*, 88–89, 92
irrigation *91*
subsidence 75

Colorado River 89, 92

Columbia River (U.S./Canada)
flood of 1948 85t
volcanic eruptions 89, 156t

comets 133, *141*, 143–145, 158, 169g *see also specific comet (e.g., Halley's)*

Concepción, Chile earthquake (1939) 17t

conduit 169g

continental crust 3, 12

continental drift 169g

continental glacier 169g

continental plate 3, 12

continental shelf 63, 98, 169g

continental shield 169g

continental slope 63, 169g

continents 3–5m, 169g

convection 5–8, 42, 169g

Cook Inlet (Alaska) 56

coral 152, 155

cordillera 169g

Cordilleran ice sheet 123

core 5–7m, 46, 169g

Coso Volcanoes 50

cotton *108*

Cowlitz River (Washington) 85t, 89

Crater Lake (Oregon) 50, *81*

craters **133–148**, 135m–136m, *137–138*, 172g

Craters of the Moon National Monument (Idaho) *83*

craton 169g

creep 60

crescent dunes 115

Cretaceous period 150t, 153–156, 159–160

Crete 36

crevasse 170g

crinoids 153

crust 2m, 6m, 170g

crustal plates 1–2m, 170g

crystal 146

cyanide 48

cyclones 63

D

dams 59, 66

Daytona Beach, Florida submarine slide (1992) 63

Death Valley, California mudflow *61*

debris slide 52

Deccan Traps (India) 145

deflation 114

deforestation 65, 111–112, 162–164

delta 170g

Denver, Colorado 75

desertification x, **107–112**

deserts **102–118**, 105m, 124, 145 *see also specific desert (e.g., Mojave Desert)*

Devonian period 134, 149–150t, 152

diapirs 4, 42–42m, 170g

diatoms (single-cell algae) 125

dikes 43

dinosaurs 134–135, 153, 156, 159

divergent plate boundary 170g
dodos (extinct birds) 162
"domino effect" 164
drainage basins 98
drainage patterns 98
Drake, Colorado 89
droughts 109, 116–118, 170g
drumlin 170g
dune 170g
dust bowls 102, *103*, **116–118**
dust clouds 127, 145, 147
dust plumes 147
dust storms ix–x, **102–118**, 124, 145

E

Earth (planet) *see specific subjects* (*e.g.,* craters)
earthflows 60, *61*
earthquakes ix, 16–17t, **16–34**, 27m, 170g *see also quake site* (*e.g.,* San Francisco, California)
    damage **30–33**, 32m
    geological dynamics 2
    ground failures 70–73
    ground shifts 68, *69*
    landslides 52–55
    plate interactions 13, 21m
    resurgent calderas 80
    subduction zones 10
    subsidence 74, 76–77
East African Rift Valley 14, 21, 45
Eastern, U.S. 85t
East Pacific Rise 9, 170g
Egypt 104
El Asnam, Algeria earthquake (1980) 28
elastic rebound theory 170g

El Chichon, Mexico 41, *41*
Elm, Switzerland 58
El Niño 90m
Eocene epoch 154
eolian 170g
epicenter 170g
erosion
    coastal 98, 131
    soil x, *64*, 65m, **65–67**, 102, 107–113, *110*, *117*, 138, 140
    wind 113–114
erratic 170g
Erzincan, Turkey earthquake (1939) 17t
escarpment 170g
esker 170g
Ethiopia 156t
Etna, Mount *see* Mount Etna
Eurasian plate 3m, 13, 22, 120, 122
Europe *see also specific country* (*e.g.,* Italy); *geographic feature* (*e.g.,* Alps)
    glaciers 123, 127
    meteorites 148m
evolution 160
expansive soils 61
explosions 48–49, 55
extinctions x, 134–135, 145, 149–150t, **149–165**, 158t
extrusive 171g

F

Fairbanks, Alaska flood (1967) *86*
famines 117
faults 13m, *14*, 14–16, 20, 22, **23–27**, *25*, 27m, 28, 45, 76, 171g *see also specific fault* (*e.g.,* San Andreas Fault)
Feather River (California) *96*
feldspar 146
Fennoscandian glacier 123
Fertile Crescent 108

fertilizers 65, 67, 111
fish 162
fissures 20, 27, 46, *72*, 76, 171g
flash floods 94–95
flood control **99–101**
floodplains 84, 94, 99
floods ix–x, 84–86t, **84–101**, *87–88*, 95m *see also flood site* (*e.g.,* Big Thompson River)
    extinctions 156t
    glaciers 124–125
    ground failures 71
    landslides 53–54
    mudflows 61–62, *89*
    subsidence 76
    volcanic eruptions 40–41, 45m
Florida
    drainage basins 98
    sea levels *132*
    sinkholes 81–82, *82*
    submarine slides 63
flow failures 71–72
fluvial 171g
focus 171g
folds 28
foreshock 25
forest fires 111, *163*
formation 171g
fossils *151–152*, 154
fracture zone 8–9
France
    glaciers 124, *126*
    meteorites 143
frogs 163
frost heaving 171g
frost polygons 171g
fumaroles 48–49, 171g

G

Galveston, Texas 75, 89
Ganges River (Asia) 98, 112
genetic engineering 65
geology **1–4**
geomagnetic field reversals 128, 146, 157–158
geothermal 46, 79, 125, 171g
Germany 123–124, 146

geysers 79, 171g
glacial burst 125
glacial surge **129–131**
Glacier Peak (Washington) 50
glaciers and glaciation
x, **119–132**, 122m, *126*,
171g *see also glacier
name (e.g.,* Hubbard
Glacier)
   extinctions 152, 155,
   158
   meteorite impact 146
   volcanic eruptions 45
global warming 117–
118, 128
Gobi Desert (Asia) 105m
Gohna, India rockfall
(1893) 59
Gondwana 171g
Good Friday, Alaska
earthquake (1964) *20*,
21, 63, 69, *70*, 72–73,
74, 77, 131
Gotherburg magnetic excursion 158
graben 171g
grain 116, *117*
Grand Banks, Newfoundland 63
Gran Desierto, Mexico
115
granite 3, 171g
granitic rock 3, 13, 140
graptolites 152
gravity fault 25, 171g
Great Australian Desert
105m
Great Britain 123
Great Fireball 145
Great Salt Lake (Utah)
103
Greece 36t, 36–37
greenhouse effect 117–
118, 128
Greenland 22, 120–121,
123, 127, 131
Green River (Utah) *99*
Gros Ventre, Wyoming
landslide (1925) *53*
ground failures 68, **70–
74**, 124

groundwater 74–76, 81–
82, 172g
Guam 11
Guatemala 21, *30*, 54
Guatemala City earthquake (1976) 21, *30*, 54
Gulf of Aden *See* Aden,
Gulf of
Gulf of Alaska *see*
Alaska, Gulf of
Gulf of Mexico *see* Mexico, Gulf of
guyot 172g

H

haboobs **112–114**, 172g
Halley's Comet 143
Hawaiian Islands
   earthquakes 34
   extinctions 162
   lava tunnels 83
   submarine slide 64
   subsidence 77
   volcanic eruptions
   46–47, *47*, 64
hazardous wastes 75
Hebgen Lake, Montana
earthquake (1959) *28*,
54
Heim, Albert 58
Heimaey, Iceland volcanic eruption (1973) *45*
Herculaneum, Italy 37
Hermes (asteroid) 141
Hilo, Hawaii 34
Himalaya Mountains
(Asia) 13, 22, 59, 98,
112, 124
Hindu Kush Range
(Asia) 22
Hoba West (meteorite)
143
Hole in the Ground (Oregon) 49
Homer 36
Homo erectus 107
Hood, Mount *see*
Mount Hood
Hoover Dam *101*
horn 172g
horst 172g
hot spot 46–47, 79, 172g

hot springs 79
Hubbard Glacier
(Alaska/Yukon) 131
Hudson Bay 123
Huron, Lake *see* Lake
Huron
hurricanes 63, 97
hydrogen 46m
hydrogen bomb *147*
hydrogen sulfide 48
hydrologic cycle 90, **91–
92**
hydrologic mapping **92–
94**
hydrothermal 172g
hypocenter 172g

I

Iapetus Sea 172g
ice age 123m, 125t, 172g
   craters 135
   dust storms 102
   extinctions 155, 158
   glaciers 121–129, 131
   ground failures 70
   sinkholes 82
   subsidence 78
   volcanoes 35
icebergs 91, *128*, 129,
172g
ice caps 91, **119–122**,
126, 150, 172g
ice clouds 127
ice jams 93
Iceland *45*, 46, 125
Idaho
   floods 87
   lava tunnels *83*
igneous rock 172g
ignimbrite 172g
Illinois 30
impact craters *see* craters
Imperial Valley, California earthquake (1979)
72
India
   earthquakes 18, 22
   floods 98, 112
   geological dynamics
   3m, 13
   rockslides 59

volcanic eruptions 156t
Indian plate 3m, 13
Indo-Australian plate 3m, 13
Indochina Peninsula 22
Indonesia
  mudflows 62
  resurgent calderas 80
  volcanic eruptions 35t–36t, 37–38m, 44
interglacial 131–132, 172g
intrusive 172g
Iowa 89
Iran 3m, 17t, 19, 22, 105m
Iranian Desert 105m
Iranian earthquake (1990) 17t, 19
Iranian plate 3m
Iraq 108
iron oxides 114
irrigation 65–66, 78, 108–109m, 110, 117–118
island arcs 172g
  earthquakes 21–22
  geological dynamics 2
  plate interactions 14
  subduction zones 10–11m, 21–22
  volcanoes 21, 43
Italy 124
  avalanches 58
  calderas 55
  earthquakes 16t
  subsidence 76
  volcanic eruptions 36–37, 55
Ivory Coast 146

J

James River, South Dakota floods (1969) 85t
Japan
  earthquakes 16t, 18, 34, 69, 73
  geological dynamics 10–11

ground failures 69, 73
meteorites 144
subsidence 76
Japan, Sea of 11
Java 44, 62
Java Trench 44
Jefferson, Mount *see* Mount Jefferson
Johnstown, Pennsylvania
  flood of 1889 86–87
  flood of 1977 85t
Joshua Tree, California 24
Juan de Fuca plate 3m, 13
Jurassic period 153

K

Kalahari Desert 105m
Kansas
  floods 84t–85t
  meteorites 144
Kansu, China earthquake (1920) 16t, 18, 60, 72
karst 82, 173g
Katmai, Mount *see* Mount Katmai
Kelut Volcano, Java 62
Kentucky River flood (1977) 85t, 92
kettle 173g
Kilauea, Mount *see* Mount Kilauea
Kodiak, Alaska 21
Krakatoa, Indonesia volcanic eruption (1883) 36t, 38, 44
Kwanto Plain, Japan earthquake (1923) 16t, 18

L

laccolith 43
lahars 62, 173g
Lake Chad (Nigeria/Niger/Chad) 103
Lake Huron (U.S./Canada) 134
Lake Mamoum (Cameroon) 48

Lake Mead (Nevada/Arizona) *101*
Lake Nios (Cameroon) 48
lamellae 146
Landers, California earthquake (1992) 24
landslides ix, **52–57**, *53, 57, 62*, 173g
  dust storms 116
  earthquakes 16, 19–20, 27, 69
  rockslides 57–59
  soil slides 60–63
  submarine slides 63–65
  volcanoes 40, 48
lapilli 173g
La Soufrière, Mount *see* Mount La Soufrière
Lassen, Mount *see* Mount Lassen
Las Vegas, Nevada 76
lateral spreads 71–72, *72*
Laurasia 173g
Laurentide glacier 123
lava 2, *45*, 46–48, 83, 157, 173g
lava tunnels 83
Lexell's Comet 143
limestone 81–82
linear dunes 115
liquefaction 70–73, 173g
Lisbon, Portugal earthquake (1775) *18*, 19, 34
lithosphere and lithospheric plates 173g
  earthquakes 21–22
  geological dynamics 1–3m
  mantle convection 4–8
  seafloor spreading 10
  volcanoes 42m, 42–44
Little Conemaugh River (Pennsylvania) 86
Little Ice Age 117, 127
Lituya Bay, Alaska rockslide (1958) 59

loess 173g
Loma Prieta, California
earthquake (1989) 26,
26, 31, 54, 71
Longarone, Italy ava-
lanche (1963) 58
Long Beach, California
subsidence 75
Long Valley Caldera
(California) 50, *80m*
Los Angeles, California
earthquakes 24–26
floods 85t
landslides 54
Louisiana
earthquakes 30
floods 95, *100*

M
Madagascar 14
Maderia, Azores 34
Madison Canyon, Mon-
tana landslide (1959)
55
magma 173g
geological dynamics
2–4
mantle convection 6
plate interactions 14
resurgent calderas
79–81
seafloor spreading 8–
9
subduction zones 9–
11
volcanoes 36, 38,
42m, 42–44
magnesium oxide 114
magnetic poles 146, 158
magnetic reversals 146,
158–158t
magnitude scale 18,
173g
Maharashtra, India
earthquake (1993) 18
mammals 154–155, 162
Mamoum, Lake *see*
Lake Mamoum
Managua, Nicaragua
earthquake (1972) 21
Manicouagan structure
134, 136m

mantle 173g
convection **5–8**
geological dynamics
2m, 4–7m
plumes 5, 7, 43, 46,
79
seafloor spreading 10
volcanoes 42m, 42–
46
Mariana Trench 11
Mars (planet) *107*, 137,
143
Martinique 38–39, *39*
Maryland *120*
Mascati, Sicily 37
Massachusetts 30
mass extinctions *see* extinctions
mass wasting 173g
Mauna Loa Volcano
(Hawaii) 47, *47*, 64
Mauritania 109
Mazama, Mount *see*
Mount Mazama
McLoughlin, Mount *see*
Mount McLoughlin
McMurdo Sound 106
Mead, Lake *see* Lake
Mead
Medicine Lake Volcano
50
Mediterranean Sea 22,
107
megaplumes 44, 173g
meltwater *see* glacial
blast
Mercury (planet) *138*
Mer de Glace Glacier
(France/Switzerland)
126
Mesozoic era 120, 154
Messina, Italy earth-
quake (1908) 16t
metamorphic rocks 3,
13
meteoritic crater *134*,
174g
meteoritics 174g
meteoroids 140, 143,
174g
meteors and meteorites
x, 128, **133–148**, *134*,

135–136m, *137–138*,
139m, 144m, 146m,
158t, 173g
meteor showers 143,
146, 174g
Mexico
earthquakes 21, 70
ground failures 70
sand dunes *115*
volcanic eruptions
41, *41*, 49, 55
Mexico, Gulf of 49, 66,
98, 124
Mexico City earthquake
(1985) 21, 70
microearthquake 174g
micrometeorites 174g
microtektites 174g
Mid-Atlantic Ridge 4, 9,
45, 64, 174g
Middle East 108 *see
also specific country
(e.g., Syria)*
midocean ridge 2, 4,
174g
Midwest, U.S. 82, 85t–
86t, 90
Milankovich model
127m
Milwaukee, Wisconsin
30
minerals *see specific
mineral (e.g.,* quartz)
Minoan civilization 36
Mississippi 84t–85t, 95
Mississippian period
150t
Mississippi River
dust storms 118
earthquakes 20, 27
floods 92, 95, 98,
118, 124
glaciers 124
ground failures 68
soil erosion 66
Missouri
earthquakes 15, 19–
20, 68, 77
floods 84t–85t
ground failure 68
subsidence 77

Mojave Desert (California) 25, 103
Mono-Inyo Craters (California) 50
monsoons 59, 112
Montana *28*, 54, *55*
Monte Toc (Italy) 58
moon 8, *134*, 136
moraines 125, *126*, 174g
Mosquito Lagoon (Florida) *132*
Mount Adams (Washington) *4*, 50
mountain roots 174g
mountains 3–4, 7, 12–13 *see also specific mountain range (e.g., Himalaya); mountain (e.g., Mount Etna)*
Mount Augustine (Washington) 44
Mount Baker (Washington) 50–51
Mount Etna (Italy) 37, 55
Mount Hood (Oregon) 50–51
Mount Jefferson (Oregon) 50
Mount Katmai (Alaska) 44, 48
Mount Kilauea (Hawaii) 46
Mount La Soufrière (St. Vincent) 39
Mount Lassen (California) *50*, 51
Mount Mazama (Oregon) 81
Mount McLoughlin (Oregon) 50
Mount Pelée (Martinique) 38, *39*
Mount Pinatubo (Philippines) 41, 44
Mount Rainier (Washington) *4*, 50–51
Mount St. Augustine (Alaska) 56
Mount St. Helens (Washington) *12*, 13, 39–40,
40, 44, 48–49, 51, 62, *78*, 89, 95, *157*
Mount Shasta (California) 50
Mount Vesuvius (Italy) 36
mudflows 40–41, 55, **61–63**, *62*, 95

N

Namib Desert 105m, *114*
Namibia, South Africa 105m, *114*
Naples, Italy 37
Nazca plate 3m, 22, 44
Near East 22
Nebraska 117, 136
Nevada
    calderas 79–80
    extinctions *151*
    floods 76, *101*
    subsidence 76
Nevado de Colima Volcano (Mexico) 55
Nevado del Ruiz, Colombia mudflow (1985) 41, 63
Newberry Volcano 50
Newdale, Idaho 87
New England
    floods (1938) 84t–85t
    glaciers 123
New Jersey 85t
New Madrid, Missouri earthquakes (1911-1912) 15, 19–20, 68, 77
New Madrid Fault 15, 23, 27
New Mexico
    calderas 79
    craters 145
    lava tunnels 83
    subsidence 76
    volcanoes 50
New Orleans, Louisiana 30
New Quebec Crater (Canada) 136
New York State 85t
New Zealand 124
Nicaragua 21
nickel 134

Niger 103
Nigeria 103
Niigata, Japan earthquake (1964) 69, 73
Nile River (Africa) 107
1989 FC (asteroid) 142m
Nios, Lake *see* Lake Nios
nitrogen 46m
nitrogen oxide 128, 148
Nome River Valley (Alaska) *60*
normal fault 174g
North Africa *see* Africa
North America *see also specific country (e.g., Canada); geographic feature (e.g., Mount St. Helens)*
    deserts 105m
    extinctions 156t, 162
    glaciers 122–123, 130
    impact cratering 134
North American Desert 105m
North American plate 3m, 120
North Dakota 59
Northeast, U.S. 82, 85t
North Pole 120–121
Northridge, California earthquake (1994) 25
Norway 59
nu'ée ardente 174g
Nunziata, Sicily 37
nutation 127

O

Oahu, Hawaiian Islands 64
oblique fault 26
oceanic crust 2, 4, 8, 12, 14, 42, 44
oceanic plate 3
oceans *see specific ocean name (e.g. Pacific Ocean)*
Ohio
    earthquakes 27
    floods 84t–85t
    subsidence 75
Ohio River 27

oil spills 75, *161*
Oligocene epoch 155
Olympic Mountains (Washington) 31
Ontario, Canada 134–135m
Ordovician period 149–150t, 152
Oregon
    geological dynamics 3, *4*, 13
    lava tunnels 83
    meteorites 144
    mudflows 62
    resergent calderas 81
    volcanic eruptions 40, 44, 49–51
orogeny 174g
outgassing 174g
overthrust 174g
oxygen 46m
ozone layer 41, 157–158

P

Pacific Coast 85t
Pacific Ocean and Basin 2, 4, 9–10, 21, 34, 43, 47
Pacific plate 3m, 4, 23, 44, 46
pahoehoe lava 174g
Pakistan 17t
Paleocene period 150t
Paleozoic era 149–150t, 152, 160
Pangaea 4–5m, 175g
Panthalassa 175g
parabolic dunes 115
passenger pigeons (extinct birds) 162
Passumpsic River (Vermont) *93*
Patagonian Desert 105m
Pearl River, Mississippi/Louisiana flood (1978) 85t, 95
Pelée, Mount *see* Mount Pelée
Pennsylvania
    floods 85t, 86–87
    soil slides *61*

Pennsylvanian period 150t
peridotite 8, 42, 175g
permafrost 74, 175g
permeability 175g
Permian period 125t, 149–150t, 152, 154
Peru
    deserts 105m
    earthquakes 17t, 22
    geological dynamics 3, 11
    landslides 57
    volcanic eruptions 44
Peruvian-Atacama Desert 105m
petroleum 74–75
Phanerozoic eon 151
Philippine plate 3m
Philippines
    earthquakes 34
    geological dynamics 3m, 10–11, 34
    volcanic eruptions 41, 44
Philippines Trench 11, 44
Phoenicians 107
Phoenix, Arizona 113
photosynthesis 125, 157–158
phytoplankton 150, 154, 158
pillow lava 175g
Pinatubo, Mount *see* Mount Pinatubo
placer 175g
plankton 154
plate tectonics 1–2m, 5, 10, **12–15**, 175g
Plato 36
Pleistocene era 121
pluton 4, 43, 175g
Poland 123
polar ice caps *see* ice caps
pollution 99
polynyas 122
Pompeii, Italy 37
porosity 175g
Port Graham, Alaska 56
Portugal *18*, 19, 34

potassium 7
Precambrian era 23m, 125t, 150–151
precipitation *see* rainfall
Prince William Sound (Alaska) *161*
Prosperity, Pennsylvania earthflow *61*
Puget Sound (Washington) 31
pumice 175g
pyroclastics 46–48, 56, 62, 175g

Q

quartz 146
Quebec, Canada 134, 136m
Quette, Pakistan earthquake (1935) 17t
quick clays 73

R

radiation 149–150t
radioactive isotopes 7
rainfall 116m
    desertification 111–112
    dust bowls 117–118
    glaciers 124
    haboobs 112–114
    landslides 55, 60
    mudflows 62
    soil erosion 65
rain forests 111, 163
Rainier, Mount *see* Mount Rainier
rain shadow zone 106
Red Canyon Fault *28*
Red River of the North (U.S./Canada) floods (1965/1975/1978) 85t, 92
Red Sea 14, 45
reef 175g
Reelfoot Lake (Tennessee) *77*
regression 175g
reptiles 153–154, 162–163
Republican River, Kansas floods (1935) 85t

resurgent calderas 79m, **79–81**, 80m, 175g
reverse fault 25
ridge crest 175g
ridges 2m, 7m, 9
Ries Crater (Germany) 146
rift 8, 44–46
rift valley 176g
rille 176g
Ring of Fire 2, 21
Rio Santa River (Peru) 57
riverine floods 96
rivers *see specific river (e.g.,* Nile River)
rockslides 52, 55, **57–60**, *59*
Rocky Mountains 53, 59
Ross Sea 129
Russell Fiord (Alaska) 131
Russia
    deserts 118
    dust storms 113
    earthquakes 22
    glaciers 123

S

Sahara Desert (North Africa) 105m, 107–109, 117
Sahel (Central Africa) 108–109m, 117
St. Augustine, Mount *see* Mount St. Augustine
St. Helens, Mount *see* Mount St. Helens
St. Pierre, Martinique 38, *39*
St. Vincent 39
salinization 110
saltation 114, 176g
San Andreas Fault (California) 13m, *14,* 14–15, 20, 22–26, 28
sand boils 71, *72*
sand dunes **114–116**, *115*

San Fernando, California earthquake (1971) 54
San Francisco, California earthquakes (1906/1989) *19,* 20, 23, *24, 26,* 27–28, *69, 72, 73, 77*
San Francisco Peak (California) 50
Sangamon interglacial 131–132
San Joaquin Valley (California) 76
Santa Cruz Mountains (California) 26, *54*
Santa Rosa, California 20
satellites 92
Saudi Arabia
    dust storms 113
    eathquakes 14, 22
Scandinavia 22, 123
scarp 16, 27–28, 176g
Scotia plate 3m
scree (talus) 58
seabeds 82
seafloor spreading 2, 6, **8–9**, 176g
sea levels 76, 124, **131–132**
seamount 10, 176g
Sea of Japan *see* Japan, Sea of
seashore *120*, 123m, 131, 148
Seattle, Washington earthquake (c. 900) 31
sediments
    craters 145, 147
    dust storms 113–114
    extinctions 158–159
    glaciers 125
    ground failures 70–72, 74
    sinkholes 82
    soil erosion 66
    soil slides 60–61
    subsidence 78–79
seismic gap hypothesis 30

seismic waves 24, 32, **33–34**
sharks 162
Shasta, Mount *see* Mount Shasta
shatter cones 134, 139
Shenshu, China earthquake (1556) 17
Sherman Glacier (Alaska) *130*
shield 176g
shield volcano 47, 176g
shock metamorphism 139
shoreline *120*, 123m, 131
Siberia
    extinctions 156t
    glaciers 120, 124
    meteorites 144
Sicily, Italy 37
Sierra Nevada Range (California) 43
sill 43
Silurian period 149t
sinkholes **81–83**, *82,* 176g
slumps 52
smog 148
Socorro, New Mexico 50
soil erosion x, *64,* 65m, **65–67**
    craters 138, 140
    dust storms 102, 107–113, *110, 117*
soil profile 66m, *110*
soil slides 58, **60–61**
solar radiation 125, 127–128, 145, 157
Solar System *see specific planets (e.g.,* Mars); *subjects (e.g.,* meteorites)
solfatara 49
solifluction 74
solution mining 83
Somali Desert 105m
soot 111, 145, 147
South Africa
    deserts 105m, *114*
    meteorites 143

South America *see also specific country* (*e.g.,* Colombia); *geographic feature* (*e.g.,* Amazon River)
    deserts 113
    extinctions *163*
    geological dynamics 3m
    glaciers 121
    meteorites 146m
South American plate 3m, 121
South Carolina *29*, 30
South Dakota 85t, *103*
Southeast, U.S. 82, 86t
South Platte River, Colorado flood (1965) 85t
South Pole 120
Spain 34
spherules 176g
Spirit Lake (Iowa) *62*, 89
Spitak, Armenia earthquake (1988) 18
sponges 152
spreading ridge 2, 8–10, 42, 44
stable zone 22
star dunes 115
stegosaurs *153*
Stinking Water Canyon (Wyoming) *58*
stishovite 176g
stock 43
stratovolcano 48, 176g
strewn field 176g
striation 176g
subduction zones 9m, **9–11**, 11m, 176g
    earthquakes 22
    geological dynamics 2–4
    mantle convection 6
    plate interactions 14
    volcanoes 42m, 42–43, 47
submarine canyons 64, 176g
submarine slides **63–65**
subsidence 68–69, 72, **74–79**, 81, 177g
Sudan 113

Sudbury Igneous Complex 134–135m
sulfur dioxide 48
Sumbawa, Indonesia 37
Sumerians 108
sunlight *see* albedo
supercontinent 3–5m
surge glacier 177g
Switzerland 58, 124, *126*
syncline 177g
Syria 108

T

talas (scree) 58
Tambora, Indonesia volcanic eruption (1815) 35t, 37, 44
Tangshan, China earthquake (1976) 16t, *17*, 18
tectonic activity 177g
tektites 177g
temperatures 124–126, 131, 146, 157–158
Tennessee
    floods *77*
    soil erosion *64*
tephra 177g
Tertiary period 159
Tethys Sea 120, 155, 177g
Teton Dam 87
Teton River, Idaho flood (1976) 87
Texas
    erosion 131
    floods 84t–85t, 89
    subsidence 75–76
Thera, Greece volcanic eruption (1628? B.C.) 36t, 36–37m
thorium 7
Three Sisters 50
thrust fault 19, 25
thunderstorms 112–113
Tibetan Plateau 13, 22
tidal floods 97
tidal waves *see* tsunamis
till 177g
tillites 125, 177g

Toba Caldera (Sumatra) 80
Toc, Monte *see* Monte Toc
Tokyo, Japan
    earthquakes (1802/1857/1923) 18
    subsidence 76
Toutatis (asteroid) 142
Toutle River (Washington) 89
toxic gases **48–49**, 145, 157 *see also specific gas* (*e.g.,* carbon dioxide)
Transantarctic Range 106, 122
transform fault 9, 177g
transgression 177g
tree rings 103, *104*
trench 2m, 3, 9–11, 177g
Triassic period 149–150t, 153–154
trilobites 151
tsunamis (tidal waves) **33–34**, 145, 148, 177g *see also tsunami site* (*e.g.,* Niigata, Japan)
    earthquakes 19, 31, *33*
    ground failures 72
    hurricanes 97
    landslides 56–57
    meteorite impacts 145, 148
    rockslides 59
    submarine slides 63
    volcanic eruptions 36, 38, 48
tuff 177g
tundra 177g
Tunguska meteorite 144m
turbidite 177g
turbidity currents 63
Turkestan Desert 105m
Turkey 17t, 22
Turkey earthquake (1939) 17t

Turnagain Heights, Alaska landslide (1964) *70, 74*
Turtle Mountain (North Dakota) 59
turtles 154
typhoons 76

U

ultraviolet rays 158
underplating 10
United Kingdom *see* Great Britain
Unzen Volcano (Japan) 48
uranium 7
Utah
    calderas 79
    deserts 103
    floods 86t, *99*

V

Valerie Glacier (Alaska) 131
Valle del Bove caldera (Italy) 55
Valles Caldera (New Mexico) 79
Valley of Ten Thousand Smokes (Alaska) 49
Venice, Italy 76
vent animals 9
Verga 113
Vermont *93*
Vesuvius, Mount *see* Mount Vesuvius
Virginia *97*
volcanic ash 177g
volcanic bomb 177g
volcanic cone 178g
volcanic crater 178g
volcanoes and volcanism ix, 35–36t, **35–51**, 45m–46m, 50m, 178g *see also volcano*

name (e.g., Mount St. Helens)
    earthquakes 29
    extinctions 156t, 156–157, *157*, 161
    geological dynamics 2
    glaciers 125, 128
    ground failures 68, 70
    landslides 53, 55
    mantle convection 5–8
    meteorite impacts 140
    mudflows 62–63
    plate interaction 13–14
    resurgent calderas 79m, 79–81
    seafloor spreading 9
    subduction zones 2, 10–11, *78*

W

Washington (state)
    earthquakes 31
    floods 85t, 89, 95
    geological dynamics 3, *4, 12,* 13
    landslides *62*
    lava tunnels 83
    mudflows 62, *89*
    volcanic eruptions *12,* 13, 39–40, *40,* 44, 48–49, 51, 62, *78,* 89, 95
water and water vapor 46m, 55, 60, 78, 148 *see also* groundwater
water table 60, 71, 74
weather patterns *see also specific weather (e.g.,* hurricanes)
    dust storms 111, 118

glaciers 124, 127
Weddell Sea 129
Western Desert (Egypt) 104
West Indies *see specific country (e.g.,* Martinique)
wetlands 111, 131
wheat *see* grain
Willamette Metorite 144
wind erosion 113–114, 117
windstorms 114
Winslow, Arizona 135
Winter Park, Florida sinkhole (1981) 81
Wisconsin 30
Wolf Creek Crater (Australia) 136
Wyoming
    landslides *53*
    resurgent calderas 79
    rockslides *58*
    volcanic eruptions 46, 50

Y

Yakutat, Alaska 131
Yellow River, China flood (1887) 95
Yellowstone Caldera (Wyoming) 46, 50, 79
Yokohama, Japan earthquake (1923) 18
Yosemite National Park (California) 80
Younger Dryas 124
Yungay, Peru earthquake (1970) 17t, 57

Z

Zafferana Etnea, Sicily 37